ARIZONA & NEW MEXICO

FIELD GUIDE TO

BIRDS

Edited by JONATHAN ALDERFER

National Geographic
Washington, D.C.

Introduction

Arizona and New Mexico have some of the most popular bird-watching locations in North America. The Southwest has an incredible diversity of habitats, from lush cottonwood groves along the many streams and rivers, to rich cactus-laden deserts, to cool high-elevation coniferous forests. Nearly 550 species of birds have been recorded between the two states, of which about 300 nest annually within our borders.

The Sonoran Desert, easily accessed from either Phoenix or Tucson, supports several birds found nowhere else in North America, such as Gilded Flicker, Black-tailed Gnatcatcher, and Rufous-winged Sparrow. This rich habitat, dominated by huge Saguaro cactuses, is home to many regional specialties, including Gambel's Quail, Gila Woodpecker, Verdin, and Black-throated Sparrow. The isolated mountain ranges of southeastern Arizona and southwestern New Mexico are home to many other species unique to the Southwest: A summer visit to the spectacular Chiricahua Mountains, or to the lovely canyons in the Huachuca Mountains, will likely yield a number of fascinating birds, such as Elegant Trogon and Montezuma Quail; several flycatchers not found anywhere else in the United States, including Buff-breasted and Sulphur-bellied; and warblers such as Red-faced, Grace's, Olive, and Painted Redstart. In the north, the high-elevation pine forests stretching from the Grand Canyon into New Mexico, feature many Rocky Mountain species, such as Mountain Bluebird and Lewis's Woodpecker. And the agricultural valleys of the Rio Grande and Sulphur Springs attract thousands of Sandhill Cranes every winter, along with wintering sparrows and raptors.

This amazing richness brings thousands of birders to Arizona and New Mexico every year.

Gary Rosenberg
North American Birds, Co-editor for Arizona

Frontispiece: Scaled Quail
Photo by Tom Vezo

ARIZONA & NEW MEXICO

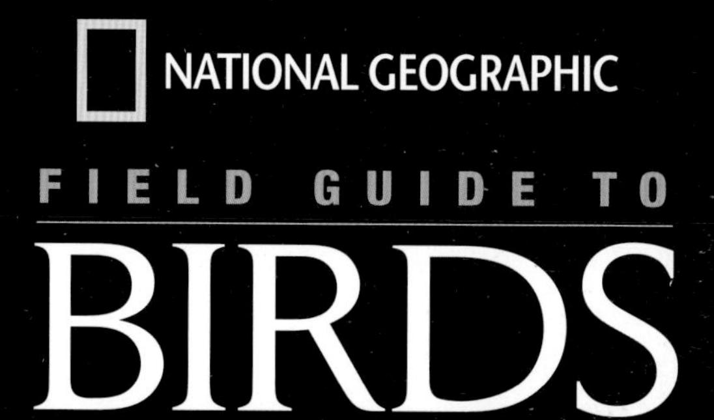

CONTENTS

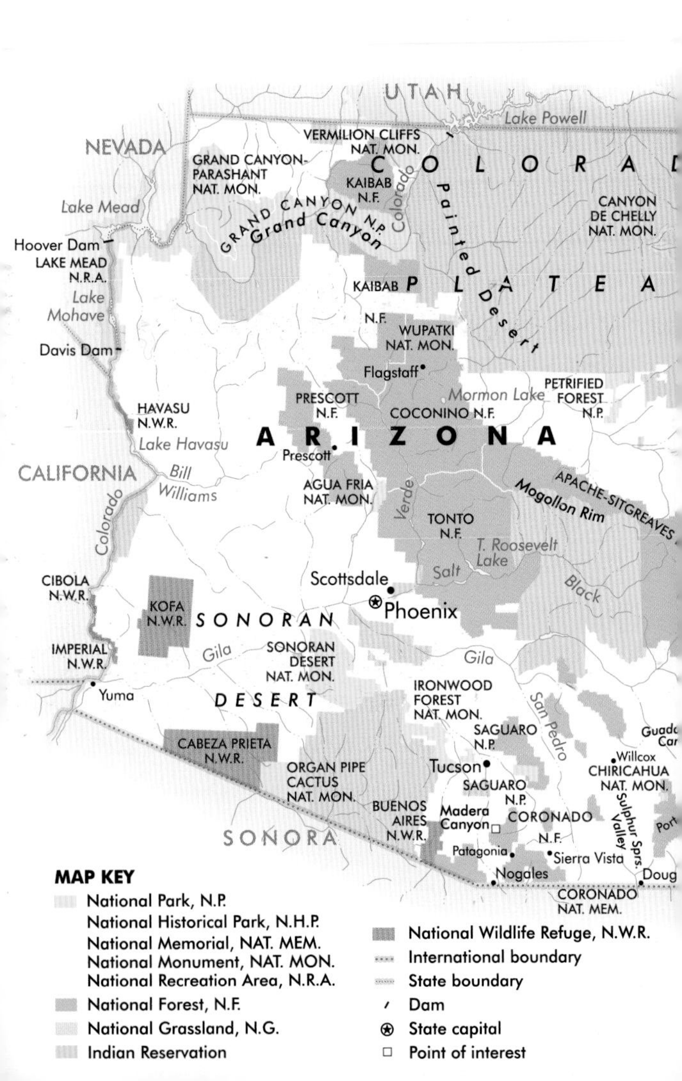
UTAH
Lake Powell
NEVADA
VERMILION CLIFFS NAT. MON.
GRAND CANYON-PARASHANT NAT. MON.
KAIBAB N.F.
COLORADO
Lake Mead
GRAND CANYON N.P.
Grand Canyon
Colorado
Painted Desert
CANYON DE CHELLY NAT. MON.
Hoover Dam
LAKE MEAD N.R.A.
Lake Mohave
KAIBAB
N.F.
PLATEAU
WUPATKI NAT. MON.
Davis Dam
Flagstaff
PETRIFIED FOREST N.P.
Mormon Lake
PRESCOTT N.F.
COCONINO N.F.
HAVASU N.W.R.
ARIZONA
Lake Havasu
Prescott
CALIFORNIA
Bill Williams
AGUA FRIA NAT. MON.
APACHE-SITGREAVES
Mogollon Rim
Verde
Colorado
TONTO N.F.
T. Roosevelt Lake
Salt
Scottsdale
Black
CIBOLA N.W.R.
KOFA N.W.R.
SONORAN
Phoenix
IMPERIAL N.W.R.
Gila
SONORAN DESERT NAT. MON.
Gila
Yuma
DESERT
IRONWOOD FOREST NAT. MON.
San Pedro
SAGUARO N.P.
CABEZA PRIETA N.W.R.
Tucson
Willcox
ORGAN PIPE CACTUS NAT. MON.
SAGUARO N.P.
CHIRICAHUA NAT. MON.
BUENOS AIRES N.W.R.
Madera Canyon
CORONADO N.F.
Sulphur Sprs. Valley
SONORA
Patagonia
Sierra Vista
Nogales
CORONADO NAT. MEM.
MAP KEY
National Park, N.P.
National Historical Park, N.H.P.
National Memorial, NAT. MEM.
National Monument, NAT. MON.
National Recreation Area, N.R.A.
National Forest, N.F.
National Grassland, N.G.
Indian Reservation
National Wildlife Refuge, N.W.R.
International boundary
State boundary
Dam
State capital
Point of interest

SELECTED BIRDING SITES OF ARIZONA AND NEW MEXICO

LOOKING AT BIRDS

What do the artist and the nature lover share? A passion for being attuned to the world and all of its complexity, via the senses. Every time you go out into the natural world, or even view it through a window, you have another opportunity to see what is there. And the more you know what you are looking at, the more you see.

Even if you are not yet a committed birder, it makes sense to take a field guide with you when you go out for a walk or a hike. Looking for and identifying birds will sharpen and heighten your perceptions, and intensify your experience. And you'll find that you notice everything else more acutely—the terrain, the season, the weather, the plant life, other animal life.

Birds are beautiful, complex animals that live everywhere around us in our towns and cities and in distant places we dream of visiting. Here in North America more than 900 species have been recorded—from abundant commoners to the rare and exotic. A comprehensive field reference like the *National Geographic Field Guide to the Birds of North America* is essential for understanding that big picture. If you are taking a spring walk in the desert Southwest, however, you may want something simpler: a guide to the birds you are most likely to see, which slips easily into a pocket.

This photographic guide is designed to provide an introduction to the common birds—and some of the specialty birds—you might encounter in Arizona and New Mexico: how to identify them, how they behave, and where to find them, with specific locations.

Discovery, observation, and identification of birds is exciting, whether you are novice or expert. I know that every time I go out to look at birds, I see more clearly and have a greater appreciation for the natural world and my own place in it.

JONATHAN ALDERFER
Editor

National Geographic Field Guide to Birds: Arizona & New Mexico* is designed to help beginning and practiced birders alike identify birds in the field and introduce them to the region's varied birdlife. The book is organized by bird families, following the order in the *Check-list to the Birds of North America,* by the American Ornithologists' Union. Families share structural characteristics, and by learning these shared characteristics early, birders can establish a basis for a lifetime of identifying birds and related family members with great accuracy—sometimes merely at a glance. (For quick reference in the field, use the color and alphabetical indexes at the back of this book.)

A family may have one member or dozens of members, or species. In this book each family is identified by its common name in English along the right-hand border of each spread. Each species is also identified in English, with its Latin genus and species—its scientific name—found directly underneath. One species is featured in each entry.

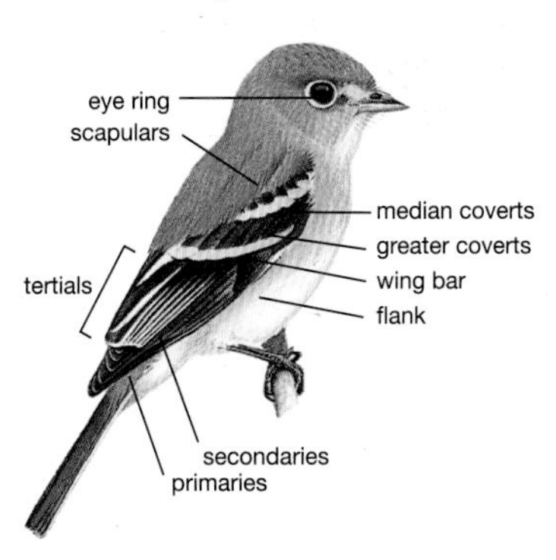

Least Flycatcher

Lark Sparrow

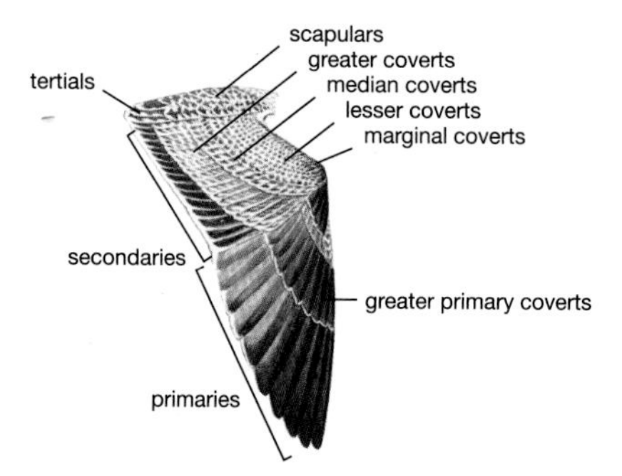

Great Black-backed Gull, upper wing

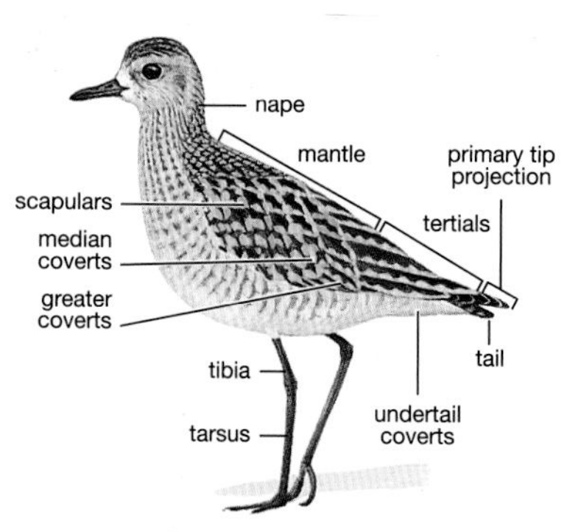

Pacific Golden-Plover

An entry begins with **Field Marks**, the physical clues used to quickly identify a bird, such as body shape and size, bill length, and plumage color or pattern. A bird's body parts yield vital clues to identification, so a birder needs to become familiar with them early on. After the first glance at body type, take note of the head shape and markings, such as stripes, eye rings, and crown markings. Bill shape and color are important as well. Note body and wing details: wing bars, color of and pattern of wing feathers at rest, and shape and markings when extended in flight. Tail shape, length, color, and banding may play a big part, too. At left are diagrams detailing the various parts of a bird—its topography—labeled with the term likely to be found in the text of this book.

The main body of each entry is divided into three categories: Behavior, Habitat, and Local Sites. The **Behavior** section details certain characteristics to look or listen for in the field. Often a bird's behavioral characteristics are very closely related to its body type and field marks, such as in the case of woodpeckers, whose chisel-shaped bills, stiff tails, strong legs, and sharp claws enable them to spend most of their lives in an upright position, braced against a tree trunk. The **Habitat** section describes areas that are most likely to support the featured species. Preferred nesting locations of breeding birds are also included in many cases. The **Local Sites** section recommends specific refuges or parks where the featured bird is likely to be found. A section called **Field Notes** finishes each entry, and includes information such as plumage variations within a species; or it may introduce another species with which the featured bird is frequently confused. In either case, an additional drawing may be included to aid in identification.

Finally, a caption underneath each of the photographs gives the season of the plumage pictured, as well as the age and gender of the bird, if discernable. A key to using this informative guide and its range maps follows on the next two pages.

Reading the Spread

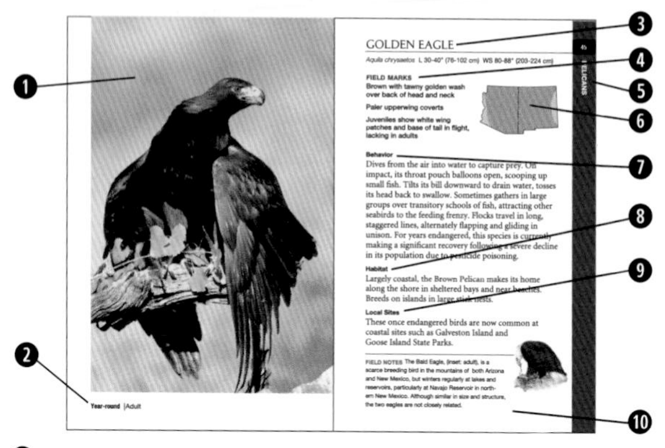

❶ **Photograph:** Shows bird in habitat. May be female or male, adult or juvenile. Plumage may be breeding, molting, nonbreeding, or year-round.

❷ **Caption:** Defines the featured bird's plumage, age, and sometimes gender, as seen in the picture.

❸ **Heading:** Beneath the common name find the Latin, or scientific, name. Beside it is the bird's length (L), and sometimes its wingspan (WS). Wingspan is given with birds often seen in flight. Female measurements are given if disparate from male.

❹ **Field Marks:** Gives basic facts for field identification: markings, head and bill shape, and body size.

❺ **Band:** Gives the common name of the bird's family.

❻ **Range Map:** Shows year-round range in purple, breeding range in red, winter range in blue. Areas through which species are likely to migrate are shown in green.

❼ **Behavior:** A step beyond **Field Marks,** gives clues to identifying a bird by its habits, such as feeding, flight pattern, courtship, nest-building, or songs and calls.

❽ **Habitat:** Reveals the area a species is most likely to inhabit, such as forests, marshes, grasslands, or urban areas. May include preferred nesting sites.

❾ **Local Sites:** Details local spots to look for the given species.

❿ **Field Notes:** A special entry that may give a unique point of identification, compare two species of the same family, compare two species from different families that are easily confused, or focus on a historic or conservation fact.

Reading the Maps

On each map of Arizona and New Mexico, range boundaries are drawn where the species ceases to be regularly seen. Nearly every species will be rare at the edges of its range. The sample map shown below explains the colors and symbols used on each map. Ranges continually expand and contract, so the map is a tool, not a rule. Range information is based on actual sightings and therefore depends upon the number of knowledgeable and active birders in each area.

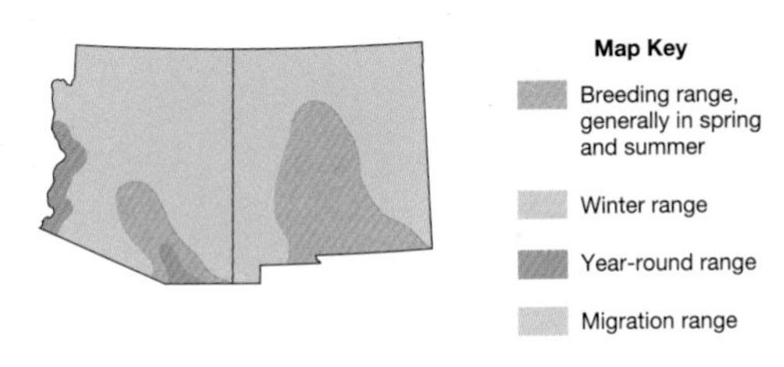

Sample Map: Green Heron

Reading the Indexes

There are two indexes at the back of this book. The first is a **Color Index** (p. 262), created to help birders quickly find an entry by noting its color in the field. In this index, male birds are labeled by their predominant color: Mostly White, Mostly Black, etc. Note that a bird may have a head of a different color than its label states. That's because its body—the part most noticeable in the field—is the color labeled.

The **Alphabetical Index** (p. 266) is organized by the bird's common name. Next to each entry is a check-off box. Most birders make lists of the birds they see. Some keep several lists, perhaps one of birds in a certain area and another of all the birds they've ever seen—a life list. Such lists enable birders to look back and remember their first sighting of an Indigo Bunting or an American Kestrel.

Year-round | Adults

BLACK-BELLIED WHISTLING-DUCK

Dendrocygna autumnalis L 21" (53 cm)

FIELD MARKS

Large duck with a long neck and long, pinkish legs

Chestnut body with black belly; large white or tan wing patch

Gray face with prominent red bill.

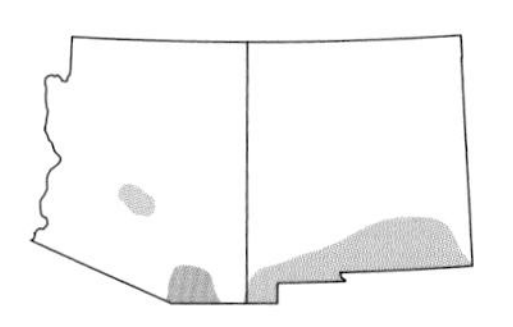

Behavior

Unlike other ducks, the Black-bellied Whistling Duck stretches out its neck and legs downward for landings and flies with slower wingbeats. In flight, feet extend beyond tail. Often perches in trees close to water. Diet consists of seeds of grasses and other plants, supplemented by small invertebrates. Forages either by day or night. Nests in tree cavities or nest boxes located near water, sometimes on the ground in dense vegetation around ponds. Pairs mate for life; both birds share nest duties and care of the young. This social, vocal bird has a unique, musical whistling call given in flight that is often repeated.

Habitat

Prefers shallow freshwater ponds and lakes, especially if bordered by trees. Nests are usually built in tree cavities or man-made nest boxes.

Local Sites

Rather uncommon breeder at ponds and along creeks and rivers in the Nogales-Patagonia area; has been found wintering at small lakes around Phoenix.

FIELD NOTES In flight, the white wing patch visible on standing birds shows as a broad white stripe set off by a black trailing edge.

Year-round | Adult white morphs

SNOW GOOSE

Chen caerulescens L 35" (89 cm) WS 45" (114 cm)

FIELD MARKS

White overall or, less commonly, dark gray-brown; black primaries show in flight

Heavy pinkish bill with black "grinning patch"

Juvenile is dingy gray-brown on head, neck, and upperparts

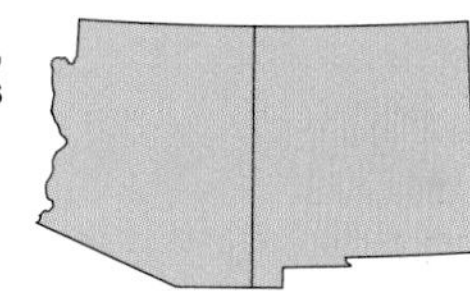

Behavior

Travels in huge flocks, especially during fall migration. Loud, vocal birds, sounding like baying hounds, flocks fly in loose V-formation and long lines, sometimes more than 1,500 miles nonstop, reaching speeds up to 40 mph. Primarily vegetarian, forages on agricultural grains and plants and on all parts of aquatic vegetation. An agile swimmer, commonly rests on water during migration and at wintering grounds. Listen for its harsh, descending *wouk,* heard continuously in flight.

Habitat

Most often seen in ponds and agricultural fields at national wildlife refuges. Breeds in the Arctic.

Local Sites

Locally abundant winter resident at refuges along the Rio Grande and Pecos River in New Mexico, such as Bosque de Apache and Bitter Lakes. In Arizona, winters at Cibola National Wildlife Refuge and, in small numbers, in Sulphur Springs Valley. Uncommon migrant elsewhere in Arizona.

FIELD NOTES The petite Ross's Goose, *Chen rossi* (inset), which can often be found in flocks of Snow Geese, has a shorter neck, rounder head, and its shorter bill lacks the "grinning patch" of the Snow Goose.

Year-round | Adult

CANADA GOOSE

Branta canadensis L 30-43" (75-108 cm) WS 59-73" (148-183 cm)

FIELD MARKS

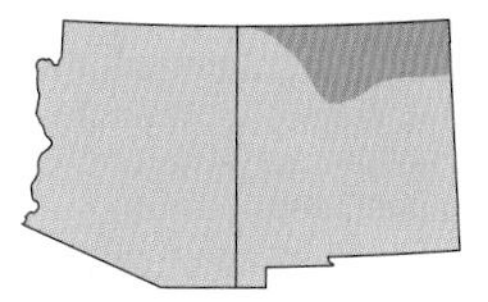

Black head and neck marked with distinctive white chin strap

In flight shows large, dark wings; white undertail coverts and a long protruding neck

Variable gray-brown breast color

Behavior

A common, familiar goose in winter; best known for migrating in large V-formation. Its distinctive musical call of *honk-a-lonk* makes it easy to identify, even without seeing it. Like some other members of its family, the Canada Goose finds a mate and remains monogamous for life. Family groups tend to stay together through the winter.

Habitat

Prefers ponds and lakes, also in cultivated fields. It has adapted successfully to habitats such as golf courses and farms, sometimes chasing off other nesting waterbirds.

Local Sites

Fairly common migrant and winter resident at most of the major wildlife refuges. Summer resident in northern New Mexico, on small ponds in the White Mountains, and on golf courses around Phoenix.

FIELD NOTES In 2004, the American Ornithologists' Union split the Canada Goose into two species, making the Cackling Goose a separate species, *Branta hutchinsii* (inset). The Cackling Goose, a rare winter visitor in Arizona and New Mexico, is smaller in size (L25-30") and has a shorter neck and stubby bill.

Breeding | Adult male

AMERICAN WIGEON

Anas americana L 19" (48 cm)

FIELD MARKS

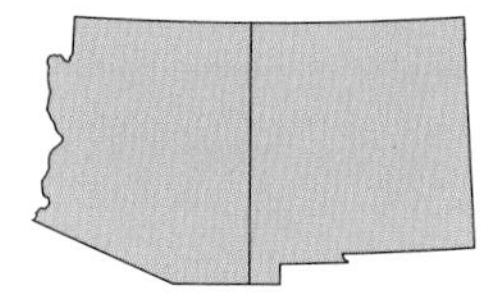

Male has conspicuous white forehead; green patch from eye to nape; large white patches in flight

Female less colorful with grayish, streaked head

Brown breast and flanks; green speculum

Behavior

A dabbling duck, largely vegetarian and prone to grazing, especially in fresh water, brackish marshes, and agricultural fields. In winter, often seen foraging on the water in mixed flocks of American Coots and diving ducks, the Wigeon is known to aggressively steal vegetation brought to the surface by the divers. Also known as the "baldpate," a name coined for the male's strikingly white forehead and crown.

Habitat

Winters throughout both states where water is available, favoring lakes and ponds. Also found at sewage ponds and golf course ponds.

Local Sites

Fairly common to locally abundant migrant and winter resident at small lakes and ponds. Rare summer breeder in extreme northern New Mexico and in ponds in the White Mountains region of Arizona.

FIELD NOTES The Gadwall, *Anas strepera*, is a common migrant and winter resident. Male (inset, right) has a black bill, blackish chest, and a plain head; female (inset, left) resembles female Mallard but has orange only on the sides of her bill.

Breeding | Adult male

MALLARD

Anas platyrhynchos L 23" (58 cm)

FIELD MARKS

Male has metallic green head and neck; white collar; chestnut breast

Female mottled brown overall; orange bill marked with black

Both sexes have bright blue speculum bordered in white; white tail and underwings

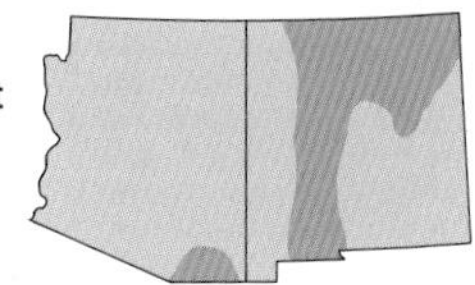

Behavior

A dabbler, the Mallard feeds by tipping into shallows and plucking seeds, grasses, or invertebrates from the bottom. Also picks insects from the water's surface. The courtship ritual of the Mallard consists of the male pumping his head, dipping his bill, and rearing up in the water to exaggerate his size. A female signals consent by duplicating the male's head-pumping. Listen for the female Mallard's loud, rasping quack.

Habitat

This widespread species occurs wherever there is shallow fresh water. "Mexican Duck" (see below) breeds along permanent streams and rivers, or at small ponds and lakes. Nests on ground in concealing vegetation.

Local Sites

Common migrant and winter resident at lakes and ponds throughout, particularly at wildlife refuges on the Rio Grande in New Mexico.

FIELD NOTES Mallards in central Mexico, formerly considered a separate species ("Mexican Duck"), lack the distincitive male plumage. "Mexican"-Mallard intergrades (inset) occur in the Southwest.

Breeding | Adult male

CINNAMON TEAL

Anas cyanoptera L 16" (41 cm)

FIELD MARKS

Male has cinnamon head, neck, and underparts

Female is mottled brown overall

Red-orange eye; long, spatulate, blackish bill

Bright blue upperwing coverts

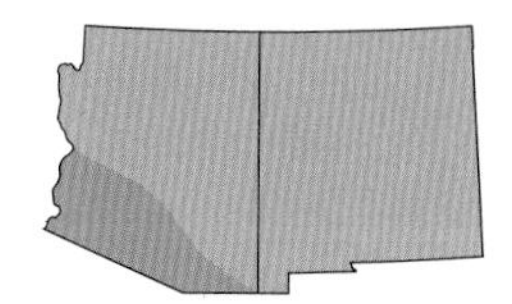

Behavior

Small but powerful, like other dabblers, the Cinnamon Teal takes flight by leaping directly into the air. Its omnivorous diet changes according to seasonal availability and particular needs during breeding, molting, and migrating. Teals typically pick insects from the water's surface or pluck grasses and invertebrates from the bottom.

Habitat

Prefers freshwater marshes, ponds, and lakes. Builds nest on ground near edges of wetlands, often below dead and matted vegetation to conceal it. The female gets into the nest through tunnels in the vegetation.

Local Sites

Fairly common to uncommon breeder on small lakes and ponds throughout Arizona and New Mexico. More commonly seen as a migrant.

FIELD NOTES In flight, wing patterns match those of the Blue-winged Teal, *Anas discors,* which is an uncommon migrant in Arizona and New Mexico. The male Blue-winged Teal (inset) has a distinctive white crescent on its face, but the female is almost identical to the female Cinnamon Teal.

Breeding | Adult male

RING-NECKED DUCK

Aythya collaris L 17" (43 cm)

FIELD MARK

Male has black head, breast, back, and tail; pale gray sides

Female is brown with pale face patch, eye ring, and eye stripe

Peaked crown; blue-gray bill with white ring and black tip

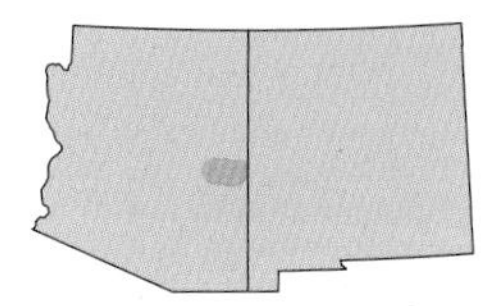

Behavior

An expert diver, the Ring-necked can feed as deep as 40 feet below water's surface, but tends to remain in shallower waters. Small flocks can be seen diving in shallow water for plants, roots, and seeds. Unlike most other diving ducks, the Ring-necked springs into flight directly from water, and flies in loose flocks with rapid wing beats. The species is named for the faint chestnut ring on its neck (visible in the photograph).

Habitat

Inhabits freshwater marshes, woodland ponds, and small lakes. Found in coastal marshes in winter. Often nests atop a floating raft of aquatic vegetation.

Local Sites

Locally common migrant and winter resident, mainly in small lakes and ponds, particularly at golf-course ponds in Phoenix and Tucson. Breeds in small numbers in the White Mountains of Arizona.

FIELD NOTES The similar Lesser Scaup, *Aythya marila,* is a fairly common migrant and winter resi-den. The male (inset, left) lacks the black back and bold white ring on his bill; the female (inset, right) has a bold white patch at the base of her bill.

Breeding | Adult male

RUDDY DUCK

Oxyura jamaicensis L 15" (38 cm)

FIELD MARKS

Breeding male has large black head with bold white cheeks; bright blue bill; rusty-red body, long black tail

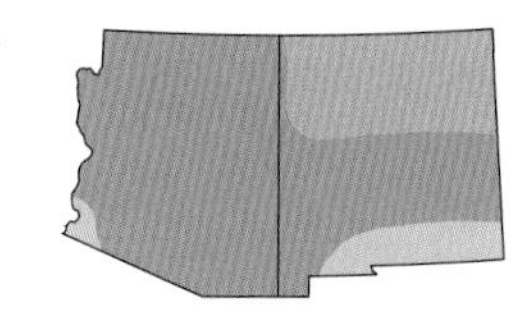

Female is dull brown overall with whitish cheek crossed by a dark line

Behavior

Referred to as a "stiff-tail" from its habit of cocking upright its long tail, which it uses as a rudder. This chunky diver is noted for its grebelike ability to sink beneath the surface and disappear. Adapted for diving, it has the largest feet relative to body size of all ducks. With legs positioned far back on its body, it can barely walk upright and rarely leaves the water. Feeds primarily on aquatic insects and crustaceans; eats little vegetable matter.

Habitat

Breeds on small lakes and ponds with marshy borders.

Local Sites

Locally common breeder along the Colorado River and on lakes and ponds, mainly above the Mogollon Rim in Arizona, and across northern New Mexico. Common to locally abundant migrant and winter resident throughout both Arizona and New Mexico.

FIELD NOTES During the breeding season, the male's bill turns from a dull blue to a bright blue. Unlike most ducks, pair bond forms upon arrival at breeding grounds and seem to last only until incubation starts.The female Ruddy Duck lays the largest eggs in relation to body size of all ducks.

Year-round | Adult male

GAMBEL'S QUAIL

Callipepla gambelli L 11" (28 cm)

FIELD MARKS

Grayish above, with prominent tear-shaped, forward-curling plume

Chestnut sides and crown

Male has dark forehead; black throat; black patch on belly

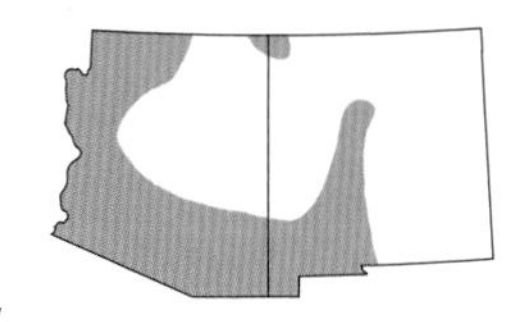

Behavior

These gregarious birds form large coveys in fall and winter. During breeding season, pairs leave the flock to nest. Calls include grunts, cackles, a plaintive *qua-el* and a loud, querulous *chi-ca-go-go,*

Habitat

Desert scrublands and thickets, usually near a permanent water source. Quail use natural depressions in the ground or scrape out shallow nests with feet and wings; may use abandoned nests of other desert birds. Nests are lined with feathers or vegetation. The female lays 9 to 15 eggs, and the hatched chicks are usually able to walk and forage within hours.

Local Sites

Common permanent resident in desert scrub, and along streams and rivers throughout southern Arizona below the Mogollon Rim, particularly in desert parks in Phoenix and Tucson, and in southwestern New Mexico north along the Rio Grande to Albuquerque.

FIELD NOTES Sometimes hybridizes with the Scaled Quail, *Callipepla squamata* (inset), which is fairly common in the higher Chihuahuan Desert, particularly around Sierra Vista, Sulphur Springs Valley and Portal areas in southeastern Arizona.

Year-round | Adult male

MONTEZUMA QUAIL

Cyrtonyx montezumae L 8¾" (22 cm)

FIELD MARKS

Plump, short-tailed, and round-winged

Male has rounded pale crest on back of head; a pied, or striped, face; spotted breast

Female is mottled pinkish-brown with less distinct head markings.

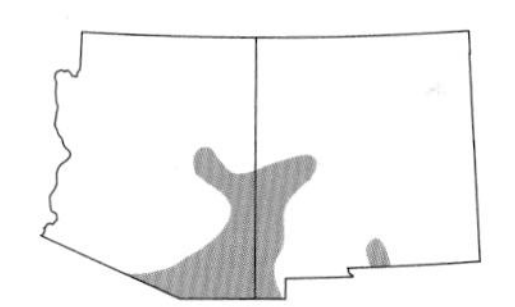

Behavior

When danger threatens, the Montezuma Quail typically crouches and remains still until the predator comes too close. Then the bird explodes into flight. During mating season, the Montezuma Quail sounds a loud, quavering, descending whistle.

Habitat

Prefers the grassy undergrowth of open juniper-oak or pine-oak woodlands on semi-arid mountain slopes. Typically builds a dome over its shallow on-the-ground nest. Female sits on clutch of 8 to 15 eggs for about 25 days.

Local Sites

This secretive quail is an uncommon permanent resident in higher elevation canyons dominated by oak in southeastern Arizona and southwestern New Mexico. Look for it along the Ruby Road west of Nogales and in Harshaw Canyon. Also occurs in the Huachucas and Chiricahuas in Arizona, and the Peloncillos and Animas in New Mexico.

FIELD NOTES More secretive and more cryptically colored than other quail, Montezuma Quail become very vocal during the late summer after the summer monsoon rains have begun.

Breeding | Adult

PIED-BILLED GREBE

Podilymbus podiceps L 13½" (34 cm)

FIELD MARKS

Short-necked; big-headed; stocky; mostly brown plumage

Breeding adult has black ring around stout, whitish bill; black chin and throat

Winter birds lose bill ring, chin becomes white

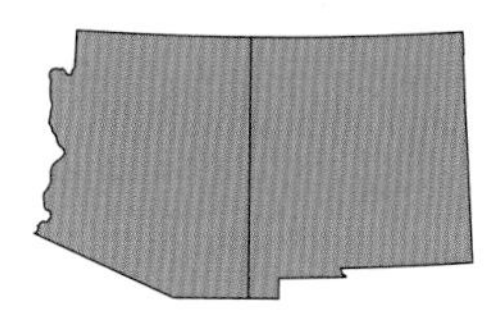

Behavior

The most secretive yet most common of North American grebes, the Pied-billed is seldom seen on land or in flight. When alarmed, it slowly sinks into the water, holding only its head above the surface. Its strong, stout bill allows it to feed on hard-shelled crustaceans, breaking apart and crushing the shells with ease. Like most grebes, it eats feathers and feeds them to its young, perhaps to protect their stomach linings from fish bones.

Habitat

Prefers nesting around lakes and ponds with marshy borders. Also along streams and rivers where cattails occur.

Local Sites

Uncommon local permanent resident at lakes and ponds throughout Arizona and New Mexico, becoming more common during migration and winter. Look for the Pied-billed Grebe wherever there are cattails, or in marsh habitats, particularly at locations such as Patagonia Lake.

FIELD NOTES In winter, the Eared Grebe, *Podiceps nigricollis* (inset), is differentiated by red eyes, thin bill, and black-and-white plumage.

Year-round | Adult

WESTERN GREBE

Aechmophorus occidentalis L 25" (64 cm)

FIELD MARKS

Long-necked with distinctive black-and-white plumage; long yellow-green bill

Black cap surrounding red eye

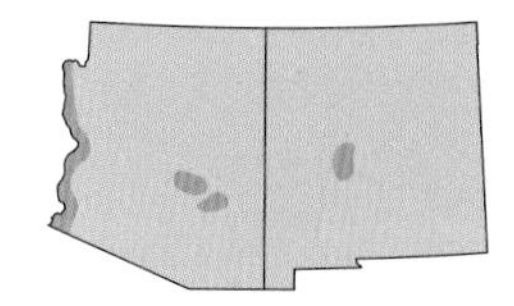

Behavior

Feeds almost exclusively on fish, diving under the surface of the water and swimming, propelled by powerful, lobed feet. Heron-like neck enables it to strike prey with a quick, snake-like strike of its long, sharp bill. Usually eats fish whole while still submerged. Elaborate courtship displays include swimming and diving together, rising up, coiling their necks, extending their wings rearward and racing across the water. Call is a loud, two-note *crick-kreek.*

Habitat

Western Grebes nest in spring in colonies among reeds along large bodies of water. Both sexes create the nest by building up mound of earth in shallow water.

Local Sites

Common breeder on large reservoirs such as Roosevelt Lake and San Carlos Lake, in Arizona; at the large lakes along the lower Colorado River, and at Elephant Butte Lake in New Mexico.

FIELD NOTES Similar to Clark's Grebe, *Aechmophorus clarkii* (inset), which also nests in both Arizona and New Mexico. Clark's Grebe has a white face that surrounds the eye. In flight, Western Grebe's white wing stripe is less extensive than Clark's.

Breeding | Adult

GREAT BLUE HERON

Ardea herodias L 46" (117 cm) WS 72" (183 cm)

FIELD MARKS

Gray-blue overall; white foreneck with black streaks; yellowish bill

Black stripe extends above eye

Breeding adult has plumes on its head, neck, and back

Juvenile has dark crown; no plumes

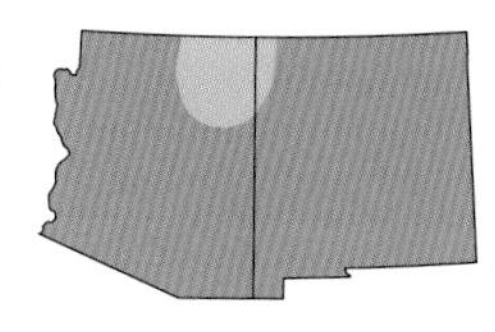

Behavior

Often seen standing or wading along calm shorelines or rivers, foraging for food. It waits for prey to come into its range, then spears it with a quick thrust of its sharp bill. Flies with its head folded back onto its shoulders in an S-curve, typical of other herons as well. When threatened, draws its neck back with plumes erect and points its bill at antagonist. Sometimes emits an annoyed, deep, guttural squawk as it takes flight.

Habitat

May be seen hunting for aquatic creatures in marshes and swamps, or for small mammals inland, in fields and forest edges. Pair builds stick nest high in trees in loose association with other Great Blue pairs.

Local Sites

Common resident near aquatic habitats throughout, rarer in the north. Isolated breeding colonies along river courses in southern Arizona and New Mexico.

FIELD NOTES Largest of the North American herons. Number of Great Blues is augmented during migration and winter by birds that have bred north of Arizona and New Mexico.

Year-round | Adult

GREEN HERON

Butorides virescens L 18" (46 cm) WS 26" (66 cm)

FIELD MARKS

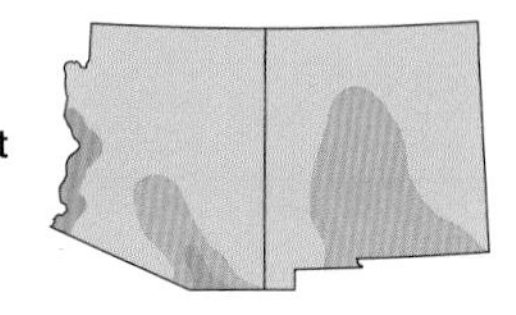

Small, chunky heron with blue-green back and crown, sometimes raised to form shaggy crest

Back and sides of neck deep chestnut, throat white

Short yellow to orange legs

Behavior

Usually a solitary hunter, a Green Heron that lands near one of its kind is likely to be attacked. Look for the bird standing motionless in or near water, waiting for a fish to come close enough for a swift attack. The Green Heron spends most of its day in the shade, but when alarmed may make a show, raising its crest, elongating its neck, and flicking its tail, in the process also revealing its lovely streaked throat plumage. This heron's common cry of *kyowk* may be heard as it flies away.

Habitat

Prefers streams, ponds, and marshes with woodland cover for breeding. Often perches in trees.

Local Sites

Uncommon and local breeder at ponds and along streams and rivers near the Colorado River, in southern Arizona, and in south-central New Mexico.

FIELD NOTES Often perches in overhanging vegetation over small ponds. In a rare instance of tool use by birds, the Green Heron is known to create its own lures to attract minnows, breaking off twigs and tossing them into the water.

Breeding | Adult

WHITE-FACED IBIS

Plegadis chihi L 23" (58 cm)

FIELD MARKS

Reddish bill; red eye; red legs

White border of red facial skin extends behind eyes, under chin

Chestnut plumage with iridescent green or purplish patches on wings and crowns; black wingtips

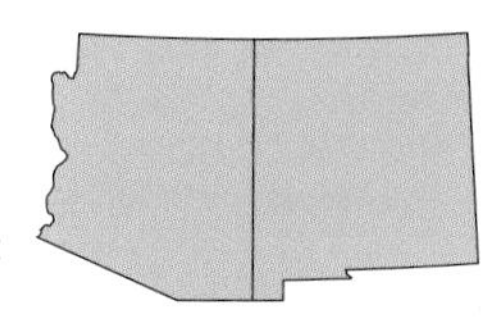

Behavior

Probes and sieves water for food with its bill. During courtship, a mating pair rubs heads together. The two offer grass and sticks to each other and preen. In flight, the White-faced Ibis appears all black. Carries neck at a downward angle, and its long legs extend well beyond its tail.

Habitat

During migration, found at lakes and reservoirs, and flooded fields in agricultural valleys. It is also commonly observed at sewage treatment ponds.

Local Sites

Common migrant wherever there is open water throughout both Arizona and New Mexico, with large concentrations sometimes found at bigger lakes and reservoirs. Has wintered in small numbers in the lower Colorado River Valley, particularly at Martinez Lake.

FIELD NOTES The Glossy Ibis, *Plegadis falcinellus,* is accidental in Arizona and New Mexico. It looks very similar to the White-faced Ibis, especially the immature birds. The adult Glossy Ibis can be distinguished by its dark facial skin surrounded by a blue border, dark (not red) eyes, and gray-green legs.

Year-round | Adult

TURKEY VULTURE

Cathartes aura L 27" (67 cm) WS 69" (175 cm)

FIELD MARKS

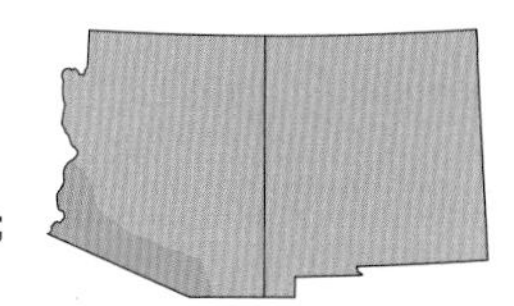

In flight,contrasting two-toned underwings; long tail extends beyond feet

Brownish black feathers on body; silver-gray flight feathers

Unfeathered red head; ivory bill; pale legs

Behavior

An adept flier, the Turkey Vulture soars high above the ground in search of carrion and refuse. Rocks from side to side in flight, seldom flapping its wings, which are held upward in a shallow V, allowing it to gain lift from conditions that would deter many other raptors. Well developed sense of smell allows the Turkey Vulture to locate carrion concealed in forest settings.

Habitat

Hunts in open country, woodlands, and farms. Often seen over highways, searching for roadkill. Found along rocky cliffs during breeding season, and in a variety of habitats during migration, including open desert and agricultural valleys.

Local Sites

Very common breeding bird and summer resident. Winters in small numbers in southwestern Arizona.

FIELD NOTES The Black Vulture, *Coragyps stratus* (inset), an uncommon resident in southwestern Arizona, is not as efficient at finding food. It will sometimes follow a Turkey Vulture to its find and claim the prey. The Black Vulture has broader wings, shorter tail, and more rapid wing beats.

Year-round | Adult male

NORTHERN HARRIER

Circus cyaneus L 17-23" (43-58 cm) WS 38-48" (97-122 cm)

FIELD MARKS

Owl-like facial disk

Slim body; long, narrow wings

Long tail with white uppertail coverts

Adult male grayish, white below with chestnut spots; female brown, buffy below with brown streaks

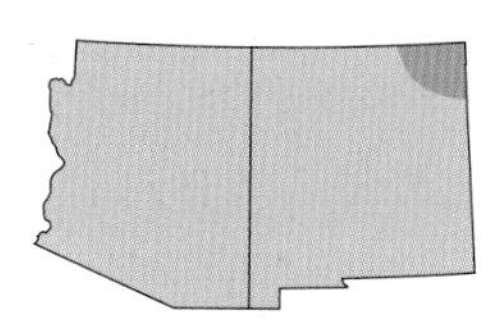

Behavior

Harriers generally perch low and fly close to the ground, wings upraised as they search for birds, mice, frogs, and other prey. Seldom soar high except during migration and in exuberant acrobatic courtship display, during which the male loops and somersaults in the air. Often found hunting at dawn or dusk, using well-developed hearing.

Habitat

Found in a variety of open habitats, including desert, grasslands, and agricultural valleys.

Local Sites

Common migrant and winter species particularly in grasslands and agricultural valleys such as the Rio Grande River Valley and the Sulphur Springs Valley south of Willcox.

FIELD NOTES Often flies slowly and low over fields looking for prey (inset: male). Look for its bright white rump. One of the only raptors—- along with the Ferruginous Hawk—likely to be seen standing in a field in winter.

Juvenile

COOPER'S HAWK

Accipiter cooperii L 14-20" (36-51 cm) WS 29-37" (74-94 cm)

FIELD MARKS

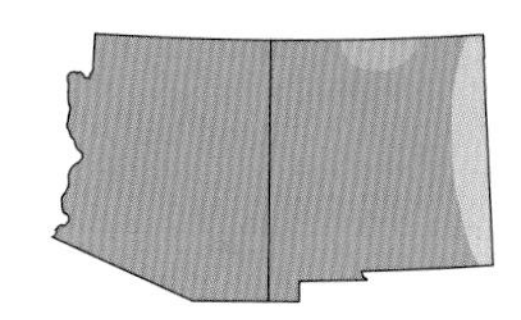

Dark gray to black cap; red eye; bue-gray upperparts

Reddish barring on underparts; long, rounded, barred tail

Juvenile brownish with streaked underparts

Behavior

Usually scans for prey from a perch, then attacks with a sudden burst of speed. Flies fast and close to the ground, using brush to conceal its rapid attack. Typically feeds on birds, rabbits, mice, squirrels, and small reptiles. Known to hold struggling prey underwater to drown it. Usually consumes prey by eating it head first, then entrails, and finally muscles.

Habitat

Found mainly along streams and rivers and in canyons. Found in a greater variety of habitats during migration and in winter.

Local Sites

Look along Sonoita Creek near Patagonia, or in any of the wetter canyons in the Santa Rita, Huachuca, or Chiricahua Mountains.

FIELD NOTES The Sharp-shinned Hawk, *Accipiter striatus* (inset: juvenile left, adult right), a very rare breeder at higher elevations and a regular migrant and winter resident, is a smaller version of the Cooper's, with a proportionally smaller head and a more squared-off tail.

Year-round | Adult

GRAY HAWK

Asturina nitida L 17" (43 cm), WS 35" (89 cm)

FIELD MARKS

Gray upperparts; gray-barred underparts and wing linings

Wings have rounded tips

Juveniles have streaked underparts; dark brown upperparts

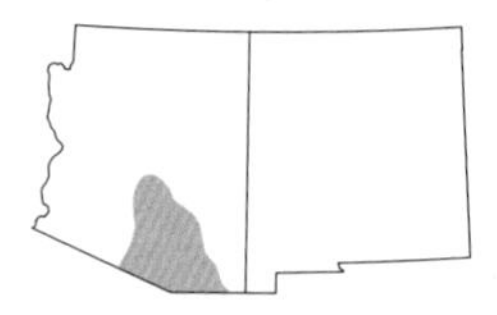

Behavior

Gray Hawks exhibit accipiter-like flight: several shallow wingbeats followed by a glide. They hunt small prey, especially lizards, in riparian woods. They frequently sound an alarm call, a very high-pitched *keeEEEErr*. Their flight call is a repeated whistle sound: *hooweeoo, hooweeoo.*

Habitat

Frequents deciduous growth along streams with nearby open land. A tropical species, the Gray Hawk is found in the United States only from Texas west through Arizona.

Local Sites

Uncommon summer resident in cottonwood-willow riparian areas of southeastern Arizona. Look along Sonoita Creek between Patagonia and Nogales, at ponds near Nogales, along the San Pedro River near Sierra Vista, and in Aravaipa Canyon.

FIELD NOTES The juvenile Gray Hawk (inset) looks quite different from the mostly gray adult. The juvenile is mostly brown above and streaked below; note its whitish face with a dark eye line and malar streak.

Year-round | Adult

HARRIS'S HAWK

Parabuteo unicinctus L 21" (53 cm), WS 46" (117 cm)

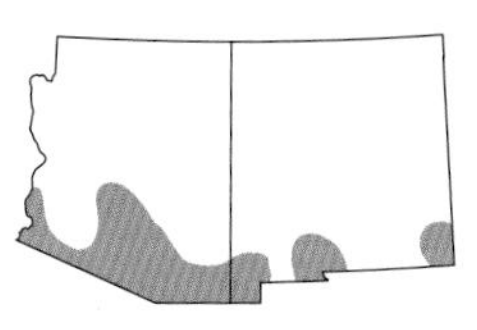

FIELD MARKS

Chocolate brown overall; conspicuous chestnut shoulder patches, leggings, and wing linings

White at base and tip of long tail

Juvenile heavily streaked below; chestnut shoulder patches less distinct

Behavior

Usually seen in loose family groups or pairs throughout the year. Often, members of a group perch a few feet apart on power poles or dead trees. These social hawks sometimes hunt in small, cooperative groups. The hunters share the food, carrying some back to nestlings. Female builds nest in small tree or saguaro cactus, and both parents incubate the two to four eggs of the clutch. The chicks leave the nest at about 38 days and can fly about 10 days later, but most stay with their extended families for years.

Habitat

Semiarid woodland and brushland replete with thorny shrubs, small trees, and prickly pear cactus.

Local Sites

Fairly common resident of southwestern New Mexico and the Sonoran Desert. Also found in the desert north of Tucson and northeast of Phoenix. In New Mexico, it may be seen around San Simone Cienega in the Animas Valley and around Laguna Grande.

FIELD NOTES The Zone-tailed Hawk, *Buteo albonotatus* (inset: adult), another raptor of the Southwest, frequents lush mountain canyons. It has a black body and wing linings; sometimes soaring on uptilted wings like a Turkey Vulture.

Year-round | Adult light morph

SWAINSON'S HAWK

Buteo swainsoni L 21"(53 cm), WS 52" (132 cm)

FIELD MARKS

Long, narrow pointed wings,

Light morph: dark bib; whitish wing linings; dark flight feathers

Dark morph: dark body; white undertail coverts, dark wings

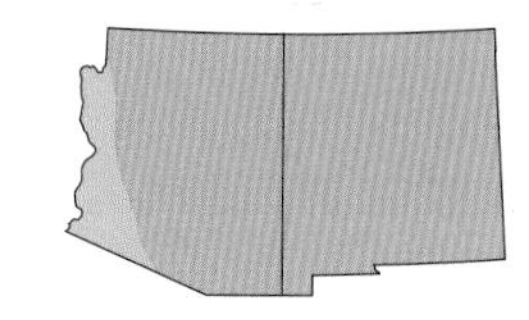

Behavior

The Swainson's Hawk soars over open plains and prairies with teetering, vulture-like flight. Its primary prey consists of large insects and small rodents, such as mice and ground squirrels. Call is a high-pitched, raspy *kreeee.*

Habitat

Open country, such as grassland, scrubland and agricultural fields. Nest is a mass of sticks lined with leafy twigs, grass, weeds, and wool, placed in a tree or on the ground.

Local Sites

A common migrant and summer resident in open grassland and agricultural fields around Sierra Vista and Sulphur Springs Valley.

FIELD NOTES Most Swainson's Hawks—including those that nest in Canada—migrate to grasslands in Argentina for the winter. Huge flocks are seen along their southbound route where the landmass narrows, such as Veracruz, Mexico. Single day counts of over 300,000 birds have been tallied.

Year-round | Adult light morph

RED-TAILED HAWK

Buteo jamaicensis L 22" (56 cm) WS 50" (127 cm)

FIELD MARKS

Brown body; heavy bill

Whitish belly with dark streaks; dark bar on leading edge of underwing

Distinctive rufous tail on adult; juvenile has gray-brown tail with numerous bars

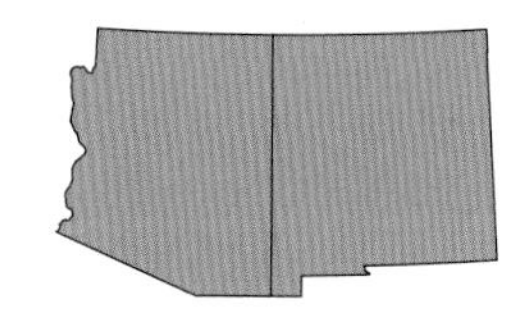

Behavior

While searching for prey, the Red-tailed Hawk hovers in place, sometimes kiting, or hanging motionless in the wind, as it scours the land. Preys on rodents. Listen for its distinctive call, a harsh, descending *keeeeeer.*

Habitat

Found in a variety of habitats including open agricultural valleys, grasslands, and open deserts. Scan for hawks along edges where fields meet desert habitats, a favored location due to the variety of prey found there.

Local Sites

A very common permanent resident throughout virtually all of Arizona and New Mexico. Numbers increase during migration and winter. The Red-tailed Hawk is the raptor most likely to be seen anywhere in the region.

FIELD NOTES The Ferruginous Hawk, *Buteo regalis* (inset: adult), a rare breeder in northern Arizona and New Mexico and a fairly common winter resident in agricultural areas in southern Arizona and New Mexico, has pale flight feathers with small dark tips and rufous "leggings."

Year-round | Adult

GOLDEN EAGLE

Aquila chrysaetos L 30-40" (76-102 cm) WS 80-88" (203-224 cm)

FIELD MARKS

Brown with tawny golden wash over back of head and neck

Paler upperwing coverts

Juveniles show white wing patches and white base of tail in flight, lacking in adults

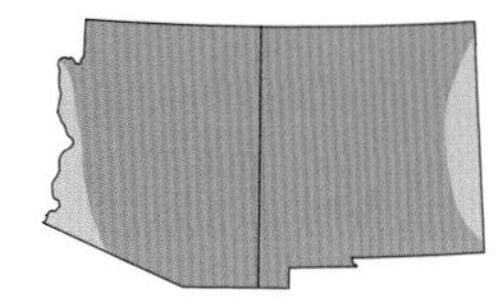

Behavior

The Golden Eagle is a true hunter and a skilled catcher of ground squirrels, jackrabbits, and waterfowl. In flight, wings are held slightly above the body, creating a slight V when viewed head-on.

Habitat

Commonly associated with mountains, canyons, and agricultural valleys in the winter. Frequently hunts over open country. Golden Eagles nest primarily on ledges and cliffs, but will use trees on occasion.

Local Sites

An uncommon but local breeder in canyons and mountains. More commonly seen during migration and winter, particularly in lowland agricultural areas such as the Sulphur Springs Valley and the Sonoita and San Rafael grasslands.

FIELD NOTES The Bald Eagle, *Haliaeetus leucocephalus* (inset: adult), is a scarce breeding bird in the mountains of both Arizona and New Mexico, but winters regularly at lakes and reservoirs, particularly at Navajo Reservoir in northern New Mexico. Although similar in size and structure, the two eagles are not closely related.

Year-round | Adult male

AMERICAN KESTREL

Falco sparverius L 10½" (27 cm) WS 23" (58 cm)

FIELD MARKS

Russet back and tail

Two black stripes on white face

Male has blue-gray wing coverts; row of white spots on trailing edge of wing

Female has russet wing coverts

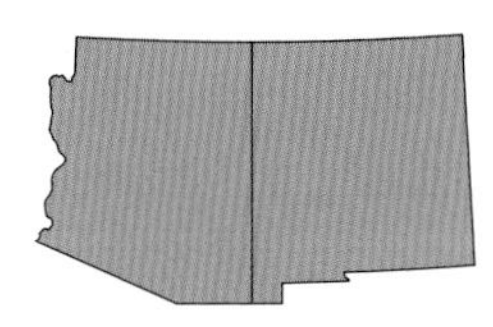

Behavior

Feeds on insects, reptiles, mice, and other small mammals, often hovering over prey before plunging. Will also feed on small birds, especially in winter. Regularly seen perched on telephone lines, frequently bobbing its tail. Has clear call of *killy killy killy.*

Habitat

The most widely distributed falcon in North America, the American Kestrel is commonly seen in open country and even in cities. Can often be found mousing along highway medians and over agricultural fields.

Local Sites

Common resident in a variety of open habitats. Numbers increase during migration and winter, particularly in agricultural areas.

FIELD NOTES The Prairie Falcon, *Falco mexicanus* (inset: adult), an uncommon, local breeding bird in much of Arizona and New Mexico, also winters in agricultural areas where American Kestrels are found. The Prairie Falcon is larger than the Kestrel and has a pale brown, rather than russet, back. In flight, note the Prairie Falcon's dark axillaries (or "wing pits") and underwing coverts, more extensive on females.

Year-round | Adult

AMERICAN COOT

Fulica americana L 15½" (39 cm)

FIELD MARKS

Blackish head and neck

Whitish bill with dark band at tip

Slate body

Outer feathers of undertail coverts white

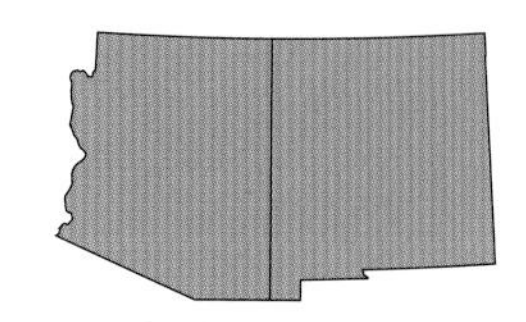

Behavior

The distinctive toes of the American Coot are flexible and lobed, permitting it to swim well in open water and even dive in pursuit of aquatic vegetation beneath the surface. A Coot has the ability to stay submerged to feed. When flushed from the water, it requires a long running start—flocks create a loud, splashy commotion.

Habitat

Nests at lakes and ponds with marshy borders. Found during migration and winter at any body of water. Has also adapted well to human-altered habitats, including pits and sewage lagoons for foraging, and golf courses for roosting.

Local Sites

Locally common breeding species at ponds and lakes throughout Arizona and New Mexico, with numbers increasing dramatically during migration and winter.

FIELD NOTES The Common Moorhen, *Gallinula chloropus* (inset), a much rarer species in the Southwest, can be distinguished from the American Coot by its red forehead shield, brownish olive back, slate underparts, and white streaking on its flanks.

Year-round | Adult

SANDHILL CRANE

Grus canadensis L 34-48" (86-122 cm) WS 73-90" (185-229 cm)

FIELD MARKS

Plumage gray overall

Dull red skin on crown and lores; juvenile lacks red patch

Whitish chin, cheek, upper throat

Flies with outstretched neck; trailing legs

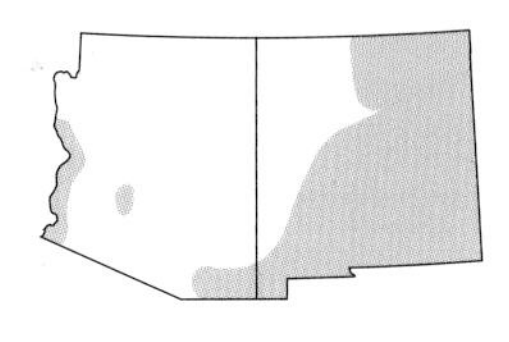

Behavior

Stands with body horizontal, picking grain, seed, fruit, insects, and small vertebrates from surface of wetlands or farm fields. Call is a loud, trumpeting *gar-oo-oo*, delivered frequently and intensely during courtship displays. Courtship consists of standardized movements of head, neck, and wings, paired with frenzied leaps resembling ballet moves. Preening with a muddy bill, the feathers of the upper back, lower neck, and breast may stain a rusty color.

Habitat

Does not breed in Arizona or New Mexico. Found in winter in agricultural valleys and national wildlife refuges.

Local Sites

Locally abundant winter resident; especially in national wildlife refuges: Bosque del Apache and Bitter Lake in New Mexico; Cibola and Havasu along the Colorado River in Arizona and in the Sulphur Springs Valley.

FIELD NOTES Both "Greater" and "Lesser" subspecies of the Sandhill Crane occur in the region. They are differentiated by size. Relocated Whooping Cranes, *Grus americana*, winter in small numbers at Bosque del Apache. They are white overall with black primaries that show in flight and a red crimson crown similar to the Sandhill's.

Year-round | Adult

KILLDEER

Charadrius vociferus L 10½" (27 cm)

FIELD MARKS

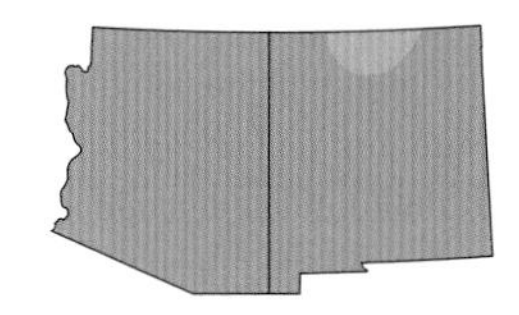

Tan to chocolate-brown above; white neck and belly

Black, double breast bands

Black stripe on forehead, extending back from black bill

Reddish eye ring

Behavior

Well known for feigning a broken wing when predators come near its nest, it will limp to one side, drag its wing, and spread its tail. Once intruders depart from the area, the instantly "healed" Killdeer will take flight, and can then be identified by its reddish orange rump. Is often seen running, stopping abruptly, then suddenly jabbing at the ground with its bill. Feeds mainly on insects found in short vegetation. May gather in loose flocks in winter. Nests on open ground, usually on gravel.

Habitat

Found along the shores of streams, lakes, and ponds. Also found in agricultural fields, particularly sod farms.

Local Sites

Locally common and widespread permanent resident along streams, rivers, lakes, reservoirs, and sewage treatment plants. It is less common in northern Arizona and New Mexico during the winter.

FIELD NOTES The Killdeer can often be identified by its by loud, piercing *kil-dee* call or its ascending *dee-dee-dee* before it is seen. Listen for a long, trilled *trrrrr* during courtship displays or when the bird is agitated.

Breeding | Adult female

WILSON'S PHALAROPE

Phalaropus tricolor L 9¼" (24 cm)

FIELD MARKS

Long, black needlelike bill

In breeding plumage, females more colorful than males; both with blackish stripe on face and neck

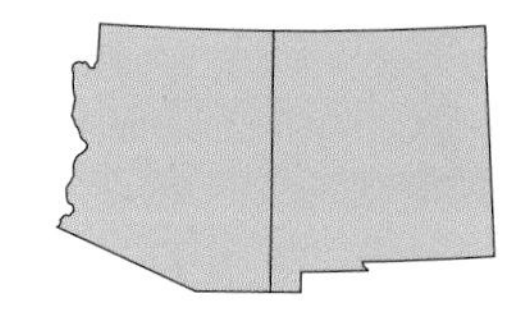

Behavior

Feeds on land and water. Foraging while swimming, the bird sometimes whirls in a circle to create a vortex that brings small prey to the surface. Also probes through mud and captures flying insects. Females, more aggressive than the males, compete with courtship displays, stretching out their necks and puffing their feathers. Females migrate first, leaving the males on the breeding grounds to incubate eggs and care for the young.

Habitat

Migrants found wherever there is open water with sandy or mud borders.

Local Sites

Common to abundant migrant at lakes and reservoirs and sewage treatment plants throughout Arizona and New Mexico. The large pond at Willco, Arizona, and any of the large reservoirs across southern New Mexico are excellent areas to view migrating phalaropes.

FIELD NOTES Winter plumage (inset), often seen during fall migration, is gray above with dark primaries and a white rump. In this plumage males and females are similar, although the females are larger.

Nonbreeding | Adult

RING-BILLED GULL

Larus delawarensis L 17½" (45 cm) WS 48" (122 cm)

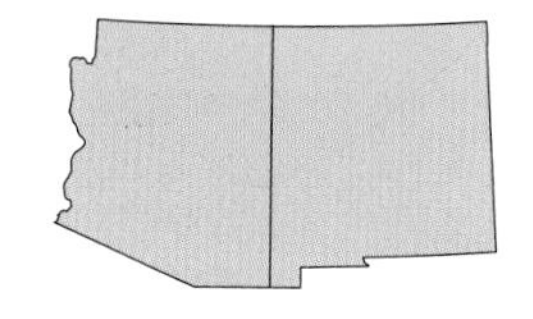

FIELD MARKS

Yellow bill with black subterminal ring

Pale eye with red orbital ring

Pale gray upperparts; white underparts

Behavior

A vocal gull often heard calling, especially during feeding and nesting. The call consists of a series of laughing croaks that begins with a short, gruff note and falls into a series of *kheeyaahhh* sounds. An opportunistic feeder, it will scavenge for seeds, grains, grasses, fruit, fish (alive or dead), marine invertebrates, and refuse.

Habitat

The Ring-billed Gull is the most frequently encountered gull at inland locations. Non-breeding birds are seen year-round.

Local Sites

Fairly common to locally abundant at lakes, reservoirs, sewage ponds, and along larger rivers during migration and in winter. Look especially along the Colorado River.

FIELD NOTES The Ring-billed Gull takes three years to reach adult plumage. The first-winter bird (inset) has a gray back, brown wings with dark, blackish-brown primaries, and a dark tipped tail. In its second winter, the gull looks much like an adult, but its bill has a broader band, it has more extensive black on its primaries, and blackish spots on the tip of its tail.

Year-round | Adult

BAND-TAILED PIGEON

Patagioenas fasciata L 14½" (37 cm)

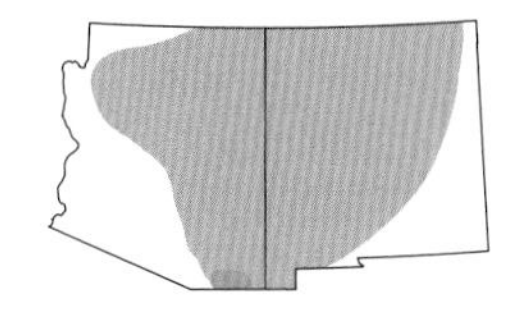

FIELD MARKS

Purplish head and breast; dark-tipped yellow bill

Broad gray terminal tail band

Narrow white band on nape of adult, bordered below by a patch of greenish iridescence

Behavior

Often perches for long spans of time either singly or in a small group at the tops of trees with little or no foliage. Size of flock may increase in winter. Forages among branches for berries, grains, seeds, nuts, and insects; rarely descends to the ground. Call is a low, repetitive *whoo-whoo,* that sounds vaguely owl-like. During breeding season, male calls from an open perch to attract a mate.

Habitat

Locally common in pine and oak woodlands. The Band-tailed is becoming increasingly common in suburban parks. Nests on platform of twigs in the fork of a tree.

Local Sites

Uncommon, but widespread summer resident. Look for it in the Jemez Mountains and near Emory Pass in New Mexico, and in the Santa Catalina and Chiricahua Mountains in Arizona.

FIELD NOTES The Rock Pigeon, *Columba livia* (inset), found mainly around cities and towns, comes in varied colorations but lacks the Band-tailed's broad gray tail band and narrow white nape band.

Year-round | Adult

WHITE-WINGED DOVE

Zenaida asiatica L 11½" (29 cm)

FIELD MARKS

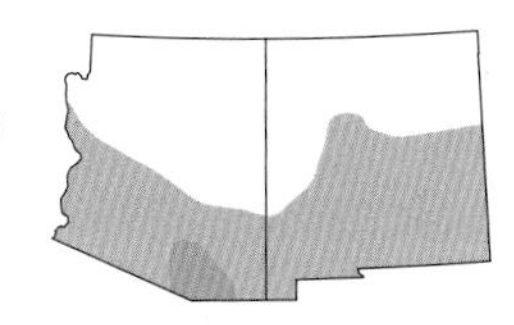

Brownish gray overall; conspicuous white wing patches

Black mark on sides of neck

Bare blue skin surrounds orange-red eyes

Rounded tail with white corners

Behavior

A desert inhabitant, will fly more than 20 miles in order to reach a source of water—whether a natural source, such as a stream, or a man-made source, such as a reservoir, canal, or cattle trough. Eats seeds, grain, and fruit. In-flight courtship display consists of male clapping his wings as he rises, then descending on stiff wings. Call is a low-pitched, drawn-out *who-cooks-for-you,* reminiscent of Barred Owl.

Habitat

Historically a bird of deserts and brushlands, the White-winged Dove is currently increasing in urban areas with scattered trees. Nests sometimes in colonies, in shrubs, mesquite, cactuses, or low in trees.

Local Sites

Common to abundant summer resident in desert habitat and riparian vegetation throughout southern Arizona and New Mexico. Very common in the Sonoran Desert and along the Rio Grande.

FIELD NOTES The White-winged Dove is expanding its range northward. Although the bulk of the population winters in Mexico, it is increasingly found in winter in towns and at feedlots in Arizona and New Mexico.

Year-round | Adult male

MOURNING DOVE

Zenaida macroura L 12" (31 cm)

FIELD MARKS

Small head with black spot on lower cheek

Trim-bodied; long tail tapering to a point

Brownish-gray upperparts; black spots on upper wings

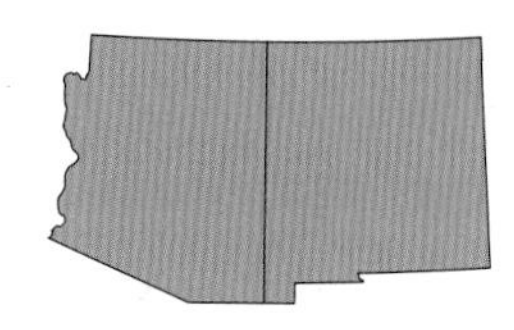

Behavior

Generally a ground feeder, a Mourning Dove will forage for grains and other seeds, grasses, and insects. Known for its mournful call, *oowooo-woo-woo-woo,* sometimes repeated several times. Wings produce a loud, fluttering whistle as the bird takes flight. A very successful breeder, a Mourning Dove may have several broods during a breeding season, each one consisting of two or three chicks.

Habitat

Found in a variety of habitats, the Mourning Dove prefers an open setting, often choosing urban or suburban sites for feeding and nesting, including front-porch eaves.

Local Sites

Common to abundant migrant and summer resident and locally abundant winter resident in southern Arizona and New Mexico.

FIELD NOTES North America's most abundant and widespread dove, and also the most widely hunted and harvested game bird. Mourning Doves produce highly nutritious "crop milk" which is secreted in the bird's crop lining. This milk, regurgitated by both parents, is the nestlings' only food for the first three days.

Year-round | Adult

INCA DOVE

Columbina inca L 8¼" (21 cm)

FIELD MARKS

Gray-buff body, paler on face

Dark edges on feathers create conspicuously scalloped pattern

Long tail edged in white

Shows chestnut primaries in flight

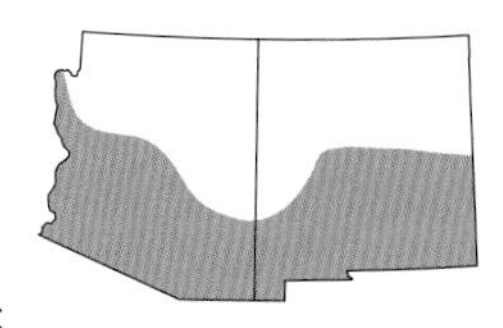

Behavior

Forages on the ground for seeds and grain, often in the company of other small doves or even of chickens on ranchlands. Ascends to perch to roost, huddled into a makeshift pyramid with other Inca Doves during colder months. A common visitor to feeders and birdbaths within its range. Male puffs out chest feathers during courtship strut. Call is a mellow *coo-coo,* repeated twice.

Habitat

Common near human habitation in semiarid regions. Female builds platform nest of twigs fairly low to ground in tree or shrub.

Local Sites

Common permanent resident, mainly in towns and cities across southern Arizona and New Mexico, though its range is spreading north and west. Very common urban bird in neighborhoods in Phoenix and Tucson, and in Socorro and Roswell in New Mexico.

FIELD NOTES The Common Ground-Dove, *Columbina passerina* (inset: male left, female right), an uncommon resident in southern Arizona and New Mexico, lacks the Inca Dove's conspicuously scalloped plumage and long tail. It prefers brushy rangelands to urban habitats.

Year-round | Adult

GREATER ROADRUNNER

Geococcyx californianus L 23" (58 cm)

FIELD MARKS

Streaked brown and white overall

Long tail edged in white

Long, heavy bill with hooked tip

Conspicuous, bushy crest

Short, rounded wings show a white crescent in flight

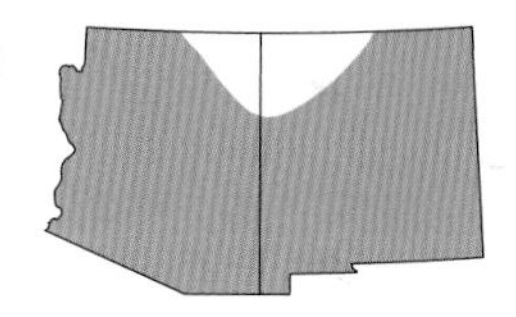

Behavior

Eats insects, lizards, snakes, rodents, and small birds; also fruit and seeds. Sprints at speeds up to 15 mph; flies only when necessary, but will glide low to the ground to gain a perch on a fence post or rock. Makes a nest of sticks lined with grass and leaves and hides the nest in cactus, shrubby bush, or a low tree. The female lays a clutch of three to six white eggs, which hatch in 20 days. Call is a dove-like *coo,* descending in pitch.

Habitat

Frequents scrub desert and mesquite groves. Also chaparral and open woodland.

Local Sites

Uncommon but widespread permanent resident found mainly in lower elevation desert habitats. Rarer and more local across the northern half of both states.

FIELD NOTES To conserve body heat on cool desert nights, the bird slows down its body functions and becomes lethargic, lowering its body temperature. A built-in heat exchanger—a patch of dark skin between its wings—helps it absorb the weak warmth of the morning sun and warm up quickly. At dawn, the bird roughs up its feathers to expose the patch.

Year-round | Adult

WESTERN SCREECH-OWL

Otus kennicottii L 8½" (22 cm)

FIELD MARKS

Gray overall; underparts marked with blackish streaks and bars

Yellow eyes; ear tufts, prominent if raised

Round, flattened facial disk

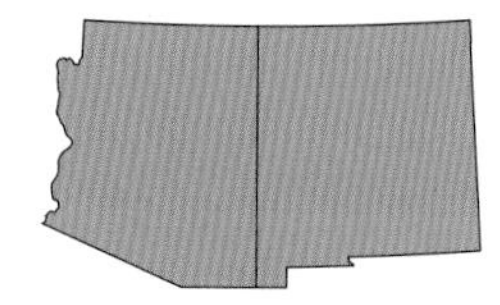

Behavior

Feeds on small rodents, songbirds, and especially insects, swooping down from a perch to seize its prey. Uneaten prey sometimes hidden in cavities to be eaten later. A nocturnal bird best identified by its call: a series of short whistles accelerating in tempo or a short trill followed by a longer trill.

Habitat

Prefers lower elevations in open woodlands, streamside groves, deserts, suburban areas, and parks. Before breeding season, the male stakes out an area with several possible nest holes—such as hollow stumps and saguaro cactus cavities—for nesting. The female then chooses the site for the nest.

Local Sites

Fairly common resident in the proper habitat across both states. Rarer and more local in the northern half of Arizona and New Mexico.

FIELD NOTES The similar Whiskered Screech-Owl, *Otus trichopsis* (inset), found in oak and pine habitats in southeastern Arizona and southwestern New Mexico, can be distinguished from the Western by its call: a series of short whistles on one pitch and at a fairly even tempo, or a series of irregular hoots, like Morse code.

Year-round | Adult

GREAT HORNED OWL

Bubo virginianus L 22" (56 cm)

FIELD MARKS

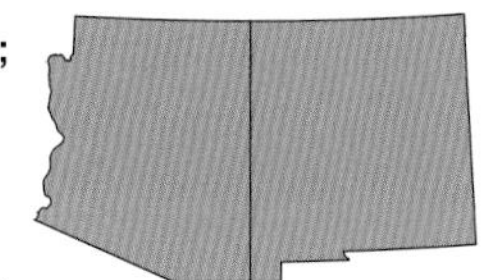

Mottled brownish gray above; densely barred below

Long ear tufts (or "horns")

Rust-colored facial disks

Yellow eyes; white chin and throat; buff-colored underwings

Behavior

Chiefly nocturnal. Feeds on a variety of animals including cats, skunks, porcupines, birds, snakes, grouse, and frogs; watches from high perch, then swoops down on prey. One of the earliest birds to nest, beginning in January or February, possibly to take advantage of winter-stressed prey. Call is a series of three to eight loud, deep hoots, the second and third often short and rapid.

Habitat

The most widespread owl in North America, the Great Horned Owl can be found in a wide variety of habitats including forests, cities, and farmlands. Reuses abandoned nests of other large birds.

Local Sites

Fairly common and widespread resident throughout both Arizona and New Mexico in a variety of forested habitats at virtually all elevations.

FIELD NOTES The Long-eared Owl, *Asio otus* (inset), a rare and local breeding bird and rare winter resident across both Arizona and New Mexico, is smaller than the Great Horned Owl and has longer, more closely set ear tufts. Its common call is one or more long *hoots*.

Year-round | Adult

ELF OWL

Micrathene whitneyi L 5¾" (15 cm)

FIELD MARKS

Tiny size; very short tail

Round head with no ear tufts

Yellow eyes with thin white eyebrows

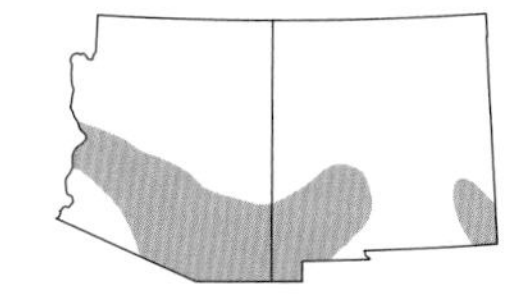

Behavior

The strictly nocturnal Elf Owl roosts in saguaro cactuses and trees, and flies somewhat like a bat. Feeds mainly on insects, catching them in the air or on the ground, but also eats mice and lizards. Call is an irregular series of high churp's and chattering notes. The female lays an average of three white eggs in a deserted woodpecker hole in a cactus or in a tree. Chicks fly at 28 to 33 days.

Habitat

Desert lowlands and canyons; especially among saguaro cactuses in the desert, and oaks and sycamores in the canyons.

Local Sites

Locally common summer resident in desert and riparian areas across much of southern Arizona and southwestern New Mexico. Most numerous in saguaro-laden Sonoran Desert. Look for it at Saguaro National Park, Sabino Canyon in Tucson, and Clanton Canyon in the Peloncillo Mountains.

FIELD NOTES Another small owl, the Northern Pygmy-Owl, *Glaucidium gnoma* (inset), is found at higher elevations. Pygmy-owl is often active during the day when its presence may first be detected by watching for mobbing songbirds. It has a long tail and distinctively streaked underparts.

Year-round | Adult

LESSER NIGHTHAWK

Chordeiles acutipennis L 8½" (22 cm)

FIELD MARKS

Long narrow wings

Cryptic plumage in brown, buff, gray, and rust

Male has white bar near tip of wing, visible in flight; in females, the bar is buff

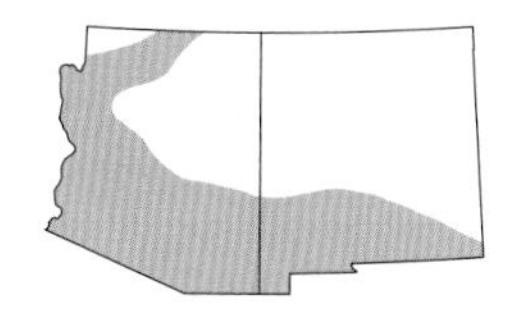

Behavior

Most often viewed at dusk. Flies low with a slow, fluttery wingbeat, feeding on night-flying insects, especially moths. Swoops down and traps the prey in its wide, bristle-edged mouth. Male courtship rituals involve stretching out plumage to show white markings. Believed to mate for life. Call a rapid, tremulous trill, heard only on breeding grounds.

Habitat

Dry, open country, scrubland, desert, valley grasslands, pastures, and prairies. Lesser Nighthawks do not use nests. Instead, the female lays two spotted, light gray eggs on bare ground and covers them with her body. Territorial males defend brooding area against intruders.

Local Sites

Common summer resident in lower elevation desert habitats across southern Arizona, particularly near sewage ponds and along the Rio Grande and the Pecos River.

FIELD NOTES The Common Poorwill, *Phalaenoptilus nuttallii* (inset), which is smaller, has broader wings, and is best identified by its song, a whistled *poor-will*. Fairly common in sagebrush or chaparral slopes; sometimes seen on roadsides at night.

Year-round | Adult

WHITE-THROATED SWIFT

Aeronautes saxatalis L 6½" (17 cm)

FIELD MARKS

White chin, breast, and center belly

White oval patch on flanks

Black upperparts

Long, forked tail; looks pointed when not spread

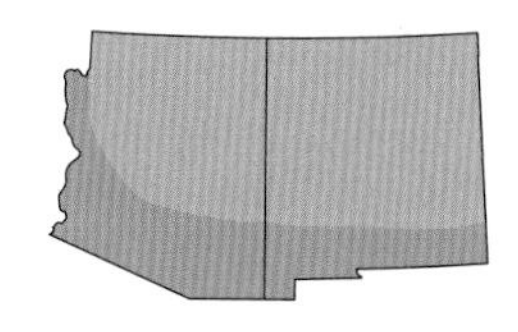

Behavior

Fast-flying bird often seen zooming around cliffs and mesas in small noisy flocks. Because of legs and feet poorly adapted to the ground, the swift does not perch during the day. It spends the time in flight, foraging for insects and ballooning spiders. Nests in crevices in cliffs, rock walls and even tall urban buildings and highway overpasses. Ranges widely during the day, traveling many miles on feeding forays. Call is a shrill chatter, *ji-ji-ji-jijiji.*

Habitat

Common near mountains, canyons, and cliffs. Because they fly so fast and forage so widely, they are also seen over desert areas and flatlands far from their nest sites.

Local Sites

Common year-round across much of southern Arizona and New Mexico; often seen foraging near ponds and lakes. Look for it at the Patagonia Roadside Rest, or along the highway up Mount Lemmon near Tucson.

FIELD NOTES The White-throated Swift is the only regularly occurring swift during the breeding season anywhere in Arizona and New Mexico. Listen for its high twittering vocalizations wherever there are high cliffs.

Year-round | Adult male

BROAD-BILLED HUMMINGBIRD

Cynanthus latirostris L 4" (10 cm)

FIELD MARKS

Mostly red bill, less in female

Male has blue gorget; blue-green underparts; white undertail coverts

Broad, forked tail is blackish-blue

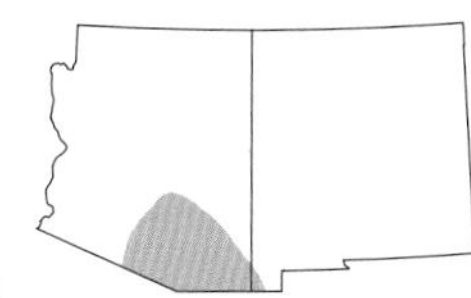

Behavior

A hummer probably consumes 1.6 to 1.7 times its body weight in flower nectar each day, and will also eat small spiders and insects. The male begins his courtship ritual by hovering about a foot from the female, then flying in repeated, pendulum-like arcs. Hummingbirds cannot walk or hop, but flight muscles are as much as 50 percent larger than those of other birds. Call is a short *je-dit*, similar to a Ruby-crowned Kinglet.

Habitat

Desert canyons, low mountain woodlands, arid scrubland, open deciduous forest. After building a nest in a small tree, the female lays two to three eggs.

Local Sites

Common summer resident in desert riparian habitat and lower montane canyons in southeastern Arizona and extreme southwestern New Mexico. Although most migrate south to Mexico in winter, the Broad-billed Hummingbird is an increasingly regular winter visitor to feeders in the Tucson area.

FIELD NOTES The paler female (inset) is quite different, with dull gray underparts, narrow white eye stripe, a dark gray ear patch, and red in the bill.

Year-round | Adult male

MAGNIFICENT HUMMINGBIRD

Eugenes fulgens L 5¼" (13 cm)

FIELD MARKS

Male: purple crown; metallic throat; black belly; tail dark green, deeply notched

Female: duller with scaly gray underparts; squarish tail with small white tips on outer feathers

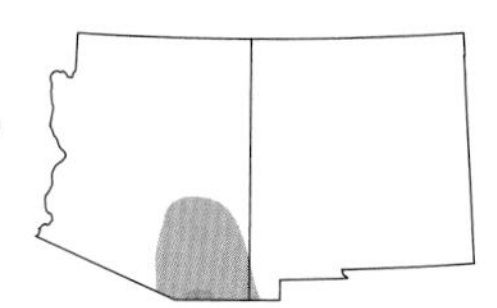

Behavior

One of our largest hummingbird, the Magnificent flies more slowly than smaller hummers; sometimes glides. Often perches prominently on high twigs. Eats more insects than other hummingbirds. Builds tiny nest of lichen and plant down bound together with spider silk on a tree limb. A clutch consists of two white eggs. The female incubates the eggs and raises the chicks alone. Main call a fairly loud *chik* or *tsik*, given from a perch and in flight.

Habitat

High mountain meadows and canyons, pine-oak woodlands, streamsides.

Local Sites

Fairly common summer resident in southeastern Arizona and southwestern New Mexico. Commonly comes to feeders at Madera, Ramsey, Miller, and Cave Creek Canyons in Arizona.

FIELD NOTES Another large U.S. hummingbird, the Blue-throated Hummingbird, *Lampornis clemenciae* (inset: male left, female right), has large white tips on the outer tail feathers and a pronounced white eyebrow. The male's throat is blue.

Year-round | Adult male

BLACK-CHINNED HUMMINGBIRD

Archilochus alexandri L 3¾" (10 cm)

FIELD MARKS

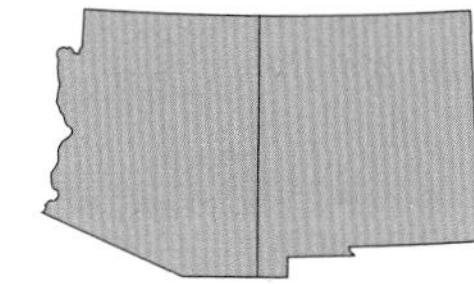

Long bill; metallic green above; whitish underparts and dusky green flanks

Male has violet band at lower border of black throat

Female's throat can be all white or show faint dusky streaks

Behavior

Twitches or pumps its tail while foraging. In cold weather, the Black-chinned Hummingbird may ingest three times its body weight in nectar in one day. Male's courtship diving display produces a zinging sound. Call is a soft *tchew*. Chase notes combine this sound with high squeals.

Habitat

A range of habitats, from urban areas to desert and from sea level to higher than 8,200 feet. Female lays two or three coffee-bean-size eggs in a camouflaged nest held together with spider silk.

Local Sites

Common migrant and summer resident in lowlands and lower montane canyons. Found at feeders in Patagonia, and at Ramsey, Miller, and Cave Creek Canyons in May and again from mid-July to mid-September.

FIELD NOTES The male Costa's Hummingbird, *Calypte costae* (inset),has a violet crown and gorget that extends down its neck. It is commonly found on brushy desert hill-sides. The female is very similar to the female Black-chinned.

Year-round | Adult male

ANNA'S HUMMINGBIRD

Calypte anna L 4" (10 cm)

FIELD MARKS

Male has deep rose-red top of head and gorget that extends onto neck

Female shows white throat speckled with red; green crown and nape

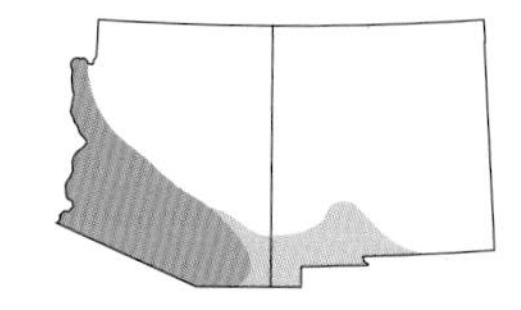

Behavior

Hovers to gather nectar, catches flying insects, and is known to pluck spiders from their webs. Common call note is a sharp *chick;* chase call a rapid, dry rattling; male's song a jumble of high squeaks and raspy notes.

Habitat

Abundant in suburban settings, mountains, and deserts, especially in winter. Somewhat nonmigratory, the adaptable Anna's Hummingbird is expanding its range northward and eastward, aided by plantings of exotic trees and shrubs.

Local Sites

Common permanent resident across much of southern Arizona, especially during migration and winter. Found increasingly in towns and cities and at feeders in Patagonia and the Huachuca Mountains.

FIELD NOTES These hummingbirds have increased dramatically as breeding birds in Arizona during the past 20 years. Females are difficult to distinguish from female Black-chinned, *Archilochus alexandri*, and Costa's, *Calpype costae,* hummingbirds. Note the shorter, straighter bill, and dark center to the throat in Anna's.

Year-round | Adult male

BROAD-TAILED HUMMINGBIRD

Selasphorus platycercus L 4" (10 cm)

FIELD MARKS

Male has green crown and rose-red gorget; mostly black tail

Female white throat with speckling; buff washed underparts; a little rufous at base of tail

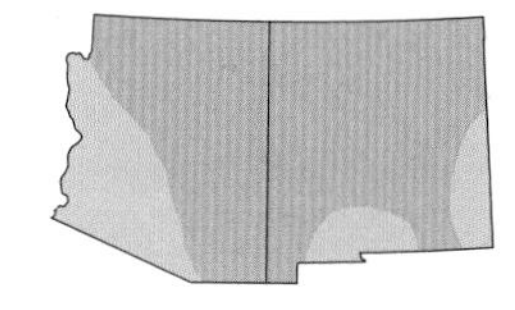

Behavior

Forages in flowers and foliage for nectar, small insects, and spiders. Except during winter molt, male's wingbeats produce a loud, metallic trill, resembling a cricket. Call is a metallic *chip.*

Habitat

Open woodlands, especially those with pinyons, junipers, conifers, and aspens; brushy hillsides and montane scrub and thickets. Males defend breeding territory. Usually nests on low horizontal tree branch, often above water. Female lays eggs in June or July and incubates them for 16 to 17 days and tends her young alone. Chicks fledge in 21 to 26 days.

Local Sites

Common summer resident at higher elevations in all of the major mountain ranges. Very common, particularly during migration, at feeders in Ramsey, Miller, and Cave Creek Canyons.

FIELD NOTES Rufous Hummingbird, *Selasphorus rufus* (inset: male), is a very common migrant, especially in southeastern Arizona. The mainly rufous male is distinctive, but the female closely resembles female Broad-tailed. Female Rufous has brighter rufous flanks and more rufous at the base of her tail.

Year-round | Adult male

ELEGANT TROGON

Trogon elegans L 12½" (32 cm)

FIELD MARKS

Stout, yellow bill; white breast-band; long, square-tipped tail

Male red below, green above

Female has red restricted to undertail coverts; brownish to grayish upperparts and head

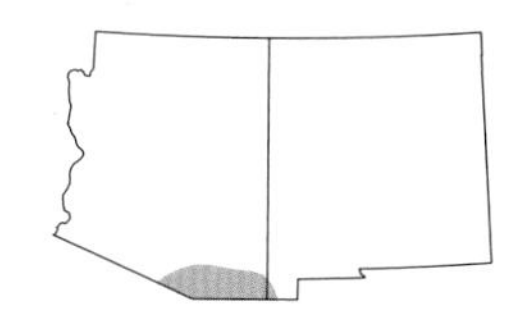

Behavior

The Elegant Trogon perches quietly with rump out and long tail pointing straight down, then tilts its head and flys out to pluck insects and small fruits. Capable of swift, evasive bursts of flight and shooting up in vertical flight. Song is a series of croaking *co-ah* notes, often the first indication of its presence.

Habitat

Streamside woodlands, mainly at altitudes of 4,000 to 6,000 feet. Female lays three or four white eggs in an unlined woodpecker hole or other cavity.

Local Sites

Uncommon and local summer resident in sycamore-lined canyons of southeastern Arizona and extreme southwestern New Mexico. Best places to look: Garden Canyon, Madera Canyon, and Cave Creek Canyon (especially the South Fork Trail).

FIELD NOTES This highly colored tropical bird is the only member of the trogon family that regularly breeds in the U.S. Similar to, but smaller than, the Eared Quetzal, *Euptilotis neoxenus,* a rare visitor from Mexico to the moutains of southeast Arizona. The Eared Quetzal has a black bill and lacks a white breast band.

Immature | Male

BELTED KINGFISHER

Ceryle alcyon L 13" (33 cm)

FIELD MARKS

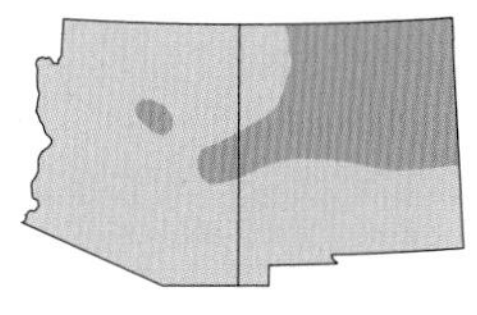

Blue-gray head with large, shaggy crest

Blue-gray upperparts and breast band; white underparts and collar

Long, heavy, black bill

Chestnut sides and belly band in female

Behavior

Generally solitary and vocal, the Belted Kingfisher dives for fish from a waterside perch or after hovering above to line up on its target. Will also forage for frogs, insects, amphibians, and small reptiles. Call is a loud, dry rattle, often given when alarmed, to demonstrate territory, or while in flight.

Habitat

Common and conspicuous along rivers, ponds, lakes, and estuaries. Prefers areas that are partially wooded.

Local Sites

Uncommon migrant and winter resident throughout much of Arizona and New Mexico wherever there is water. Commonly found along rivers such as the San Pedro and Rio Grande. Very rare as a breeding species along streams in northern Arizona and New Mexico.

FIELD NOTES The Green Kingfisher, *Chloroceryle Americana* (inset: male), has bred and is a rare, irregular winter visitor to southeastern Arizona. It is half the size of the Belted, with dark green upperparts and an inconspicuous crest. The female lacks the male's rufous breast band.

Year-round | Adult female

ACORN WOODPECKER

Melanerpes formicivorus L 9" (23 cm)

FIELD MARKS

Distinct whitish eye

Black chin, yellowish throat; white cheeks and forehead

Male has full red crown; female black in front of red

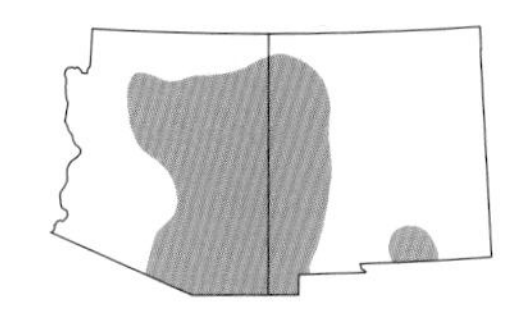

Behavior

A sociable bird, the Acorn Woodpecker can be found in small, noisy colonies, drilling holes into a tree trunk and pounding a nut into each hole for a winter supply. Will aggressively defend its store and nests from squirrels and other birds. Colony will use the same "granary tree" year after year. The bird can be seen in summer, foraging on trunks for insects or flycatching. Drinks from sap holes and feeders. Call is a repeated *waka waka waka.*

Habitat

Common in oak woods and in pine forests where oak trees are abundant. Nests communally in holes drilled into stumps or dead trees.

Local Sites

Common in sycamore-lined canyons in southeastern Arizona, such as Madera, Miller, and Cave Creek.

FIELD NOTES The related Lewis's Woodpecker, *Melanerpes lewis* (inset: juvenile left, adult right), is resident in the northern portions of Arizona and New Mexico; more widespread in winter. It is very dark, almost crow-like, and frequently flycatches from an exposed perch.

Year-round | Adult male

GILA WOODPECKER

Melanerpes uropygialis L 9¼" (24 cm)

FIELD MARKS

Neck, throat, belly, and head are grayish tan

Black-and-white barred back and rump

Male has small red cap

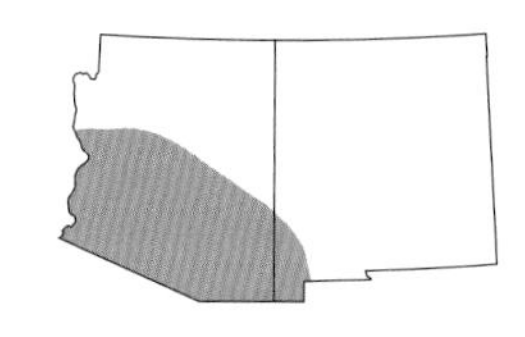

Behavior

Perches vertically, clinging to poles, cactuses, and trees, bracing itself with stiff tail feathers. Eats insects, fruits and berries. This noisy woodpecker is very conspicuous, making raucous calls as it moves about. Call is a rolling *churr* and a loud, sharp, high-pitched *yip*, often in a series.

Habitat

The Gila Woodpecker frequents towns, scrub desert, cactus country, streamside woods. Builds nests in holes made in saguaro cactuses or cottonwood trees.

Local Sites

Common permanent resident in saguaro-dominated Sonoran Desert and lowland riparian areas across southern Arizona and southwestern New Mexico. Very common in suburban areas of Tucson and Phoenix, and in well-visited birding areas such as Sonoita Creek and the San Pedro River.

FIELD NOTES Many of the holes seen in saguaro cactuses are likely made by Gila Woodpeckers. They excavate more cavities than they use, providing shelter and nest sites for many other desert species.

Year-round | Adult male

LADDER-BACKED WOODPECKER

Picoides scalaris L 7¼" (18 cm)

FIELD MARKS

Black-and-white barred back; grayish below with black spots

Face is marked by white cheek, outlined in black

Crown red on male, black on female

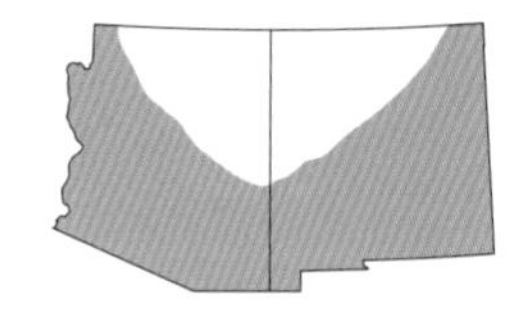

Behavior

Male and female utilize different microhabitats—male forages lower to ground, primarily for ants; female forages higher up in vegetation for other insects. Both eat the fruit of cacti. Locates stores of beetle larvae hidden beneath the bark of small trees. Known also to frequent birdbaths and feeders. Call used by male and female while foraging is a crisp, high-pitched *pik*.

Habitat

Found in arid and semiarid brushlands, as well as in mesquite and cactus country. Excavates precise, circular nesting hole in dead limbs of small trees, in large cactuses in yucca and agave plants, or in fence posts.

Local Sites

Inhabits all of the cottonwood-lined streams and rivers such as the Colorado, San Pedro, Santa Cruz, Rio Grande, and Pecos Rivers.

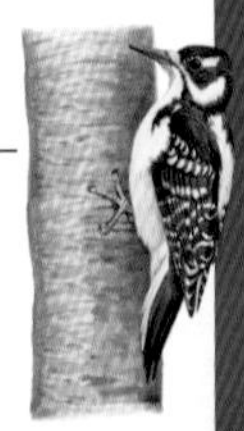

FIELD NOTES The Hairy Woodpecker, *Picoides villosus* (inset, male), found in pine forests in all major mountain ranges across Arizona and New Mexico, can be distinguished from the Ladder-backed by the unbarred white patch on its back and usually unmarked whitish flanks.

Year-round | Adult male

ARIZONA WOODPECKER

Picoides arizonae L 7½" (19 cm)

FIELD MARKS

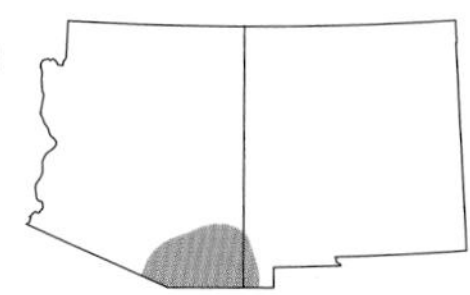

Solid brown back; underparts spotted and barred with brown

Brown crown and ear patch on white face

Male has a red hind crown

Behavior

Often forages near ground, flying from one tree to the base of another tree and working its way up the trunk into the branches. Eats larvae, adult insects, fruits, acorns. Male courtship display involves a fluttery, gliding display flight toward the female. Most common call is a sharp, hoarse *peek,* used to challenge an invasion of its territory. When disturbed, the bird sounds a rattle call—a loud, long 15-note call. Juveniles greet parents returning with food with a *tuk, tuk, tuk* call.

Habitat

The Arizona Woodpecker prefers oak or pine-oak forests or canyons. Nests are cavities dug in dead wood in evergreens, oaks, sycamores, maples, and cottonwoods. Clutch consists of two to four white eggs, which both sexes incubate for about 14 days. Chicks fledge at 24 to 27 days.

Local Sites

Most easily found in Madera, Miller, Garden, and Cave Creek Canyons in Arizona, and in the Peloncillo and Animas Mountains in New Mexico.

FIELD NOTES Formerly called "Strickland's Woodpecker," the Arizona Woodpecker is now considered a separate species from its counterpart in the mountains of central Mexico. It is the only brown-backed woodpecker in the United States.

Year-round | Adult male “Red-shafted”

NORTHERN FLICKER

Colaptes auratus L 12½" (32 cm)

FIELD MARKS

Brown, barred back; cream underparts with black spotting; black crescent bib

Brown crown and gray face

Red moustachial stripe on male, lacking on female

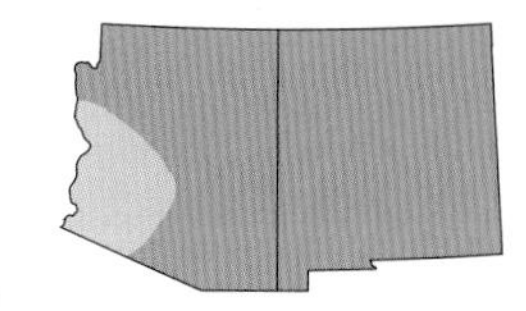

Behavior

Feeds mostly on the ground, primarily on ants, but is a cavity-nesting bird that will drill into wooden surfaces above ground, including utility poles and houses. Call is a long, loud series of *wick-er, wick-er* on breeding grounds, or a single, loud *klee-yer* year-round.

Habitat

Prefers open woodlands and suburban areas with sizeable living and dead trees. An insectivore, the Northern Flicker is partially migratory, and moves southward in the winter in pursuit of food.

Local Sites

A common migrant and winter resident in a variety of habitats, and a common summer resident of montane canyons and pine forests. Look for it on Mount Lemmon, and in the Huachuca and Chiricahua Mountains.

FIELD NOTES Once considered the same species, the Gilded Flicker, *Colaptes chrysoides* (inset: male), is restricted to the saguaro-dominated Sonoran Desert. Unlike the "Red-shafted" Northern, which has reddish underwings, the Gilded Flicker has yellow underwings and yellow at the base of its tail.

Year-round | Adult

NORTHERN BEARDLESS-TYRANNULET

Camptostoma imberbe L 4¼" (11 cm)

FIELD MARKS

Very small, drab flycatcher

Gray-olive above; very pale gray below

Bushy crest; whitish eyebrows

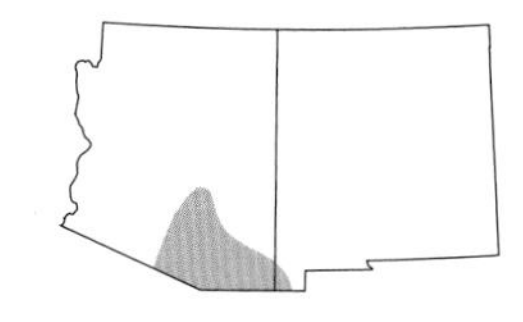

Behavior

Eats a variety of insects, using its short warblerlike bill to glean stationary insects from foliage and branches. Often moves along twigs looking for slow-moving insects. In summer, also forages by flying from its perch and catching prey in its bill. Call is an innocuous whistled *pee-yerp*. Song on breeding grounds is a descending series of loud, clear *peeer* notes—often the first indication of its presence.

Habitat

Woodland and stream thickets, stands of mesquite or cottonwood-willow in southern Arizona. Builds well-camouflaged, baseball-size nest in trees or large shrubs.

Local Sites

Fairly common but local summer resident in lowland riparian areas and desert washes in southeastern Arizona and in Guadalupe Canyon in New Mexico. Most readily seen along Sonoita Creek near Patagonia.

FIELD NOTES Another local, southeastern Arizona flycatcher, the Buff-breasted Flycatcher, *Empidonax fulvifrons* (inset), is mostly restricted to pine forests in the Huachuca and Chiricahua Mountains. The Buff-breasted Flycatcher lacks a crest, is entirely buffy-brown with two pale wing bars and a pale eye ring, and its call is a soft *pwit*.

Year-round | Adult

WESTERN WOOD-PEWEE

Contopus sordidulus L 6¼" (16 cm)

FIELD MARKS

Dark grayish brown overall; pale underparts

Broad, flat bill; yellow-orange at base of lower mandible

Long wings

Two thin white wing bars

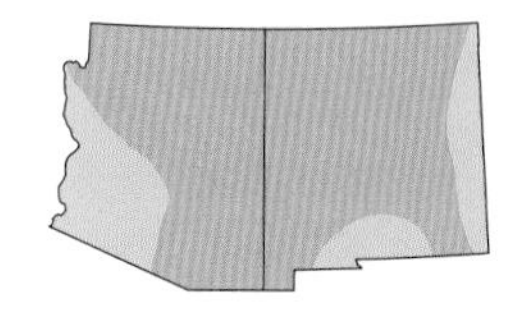

Behavior

Solitary and often hidden in trees. As it perches, the Western looks actively about, without flicking tail or wings. When prey is spotted, it darts out to catch a variety of flies, spiders, butterflies, wasps, ants, and dragonflies. Often returns to the same perch. Bristly "whiskers" help it locate prey. Calls include a harsh, slightly descending *peeer,* and clear whistles suggestive of the Eastern Wood-Pewee's *pee-yer.* Song is heard chiefly on breeding grounds and has three-note *tswee-tee-teet* phrases mixed with the *peeer* note.

Habitat

Found in open woodlands, riparian areas, in canyons, and along streams. Also in montane pine forests.

Local Sites

Common migrant and summer resident in canyons such as Garden, Madera, and Cave Creek.

FIELD NOTES The more local Greater Pewee, *Contopus pertinax* (inset), is larger than the Western Wood-Pewee and has a unique vocalization, a clear, lazy whistle that sounds like *Jose Maria*. Also note the larger bill with an orange lower mandible of Greater Pewee.

Year-round | Adult

BLACK PHOEBE

Sayornis nigricans L 6¾" (17 cm)

FIELD MARKS

Black head, breast, upperparts

Straight black bill with slight hook at tip

White belly and undertail coverts

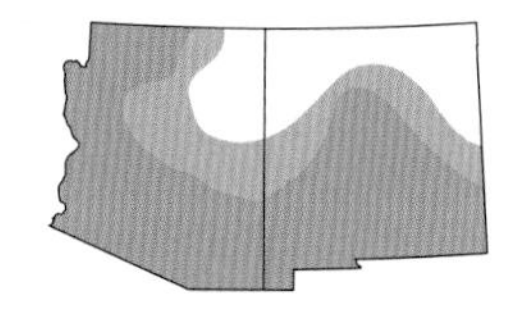

Behavior

Often seen perching upright on a low, shaded branch, pumping its tail in a distinctive downward movement. Waits to spot a flying insect, then goes after it. Frequently returns to a different perch. Has been seen to take insects and even small fish from the water's surface. Regurgitates indigestible insect parts in the form of pellets. Song is a rising *pee-wee* followed by a descending *pee-wee*. Calls include a loud *tseee* and a sharper *tsip*. Often builds nests in and on man-made structures.

Habitat

The Black Phoebe is found along streams and rivers, particularly in overhanging vegetation over small ponds.

Local Sites

Look for it along Sonoita Creek and the San Pedro River in Arizona and along the Rio Grande and the Pecos River in southern New Mexico.

FIELD NOTES Another member of the same genus, the Say's Phoebe, *Sayornis saya* (inset), is more often found in agricultural areas. It can be differentiated from the Black Phoebe by its overall brown plumage with a tawny underbelly.

Year-round | Adult male

VERMILION FLYCATCHER

Pyrocephalus rubinus L 6" (15 cm)

FIELD MARKS

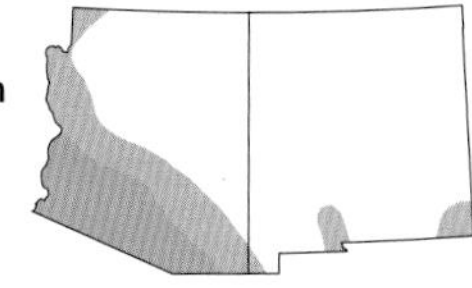

Male has striking red crown, throat, underparts; blackish brown mask and upperparts

Female has grayish brown upperparts; white breast with dusky streaking; pinkish belly

Short black tail

Behavior

Generally seen alone or in a pair, perches on low branch to scan for prey, then darts after flying insects and snags them on the wing. Also pounces on prey on the ground from a low perch. Like a phoebe, pumps and spreads its tail continuously while perched. Call is a sharp, thin *pseep.*

Habitat

Found near sources of water in arid and semiarid regions, such as streamside shrubs, small wooded ponds, and man-made irrigation zones. Female builds nest of sticks, grass, weeds, and feathers in fork of tree.

Local Sites

Comman and often conspicuous. Look for the Vermilion Flycatcher at ponds in the Nogales area of Arizona and at Rattlesnake Springs in New Mexico.

FIELD NOTES The male Vermilion Flycatcher's striking red plumage is somewhat similar to that of the Scarlet Tanager, *Piranga olivacea*, which is not found where Vermilions are found. Female Vermilion (inset) lacks the red color and is more difficult to identify. Note the slightly streaked underparts and pinkish or peach-colored lower belly in female Vermilion.

Year-round | Adult

ASH-THROATED FLYCATCHER

Myiarchus cinerascens L 8¼" (22 cm)

FIELD MARKS

Brown, bushy head; whitish gray throat

Light yellow belly; olive-brown back

Rust-colored tail feathers visible from below; rusty primaries

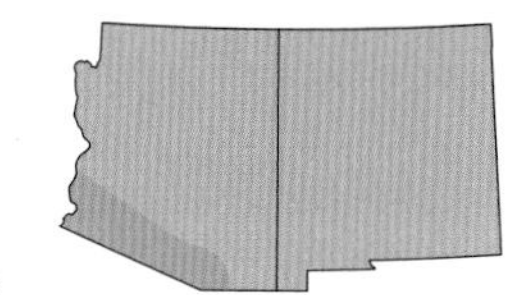

Behavior

Diet consists mainly of insects, which it usually picks from foliage. Also eats spiders, saguaro fruit, elderberries, desert mistletoe berries, and small lizards. Distinctive call, heard year around, is a burry, police whistle-like *prrt.* On breeding grounds, its song is a series of *ka-bricks.*

Habitat

Desert scrub; pinyon-juniper and oak groves; creek bottoms; dry, open woodland. Usually nests in a cavity or existing hole, which it lines with grass, hairs, weeds and twigs.

Local Sites

Fairly common summer resident in a variety of lowland and lower montane habitats across both states. Common in desert washes around Phoenix and Tucson, and along Sonoita Creek near Patagonia. Winters locally in mesquite washes in southwestern Arizona.

FIELD NOTES The Brown-crested Flycatcher, *Myiarchus tyrannulus* (inset), is very similar to the Ash-throated. The Brown-crested is a little larger with a heavier bill and is restricted more to cottonwood riparian areas. Its calls are variable but include a loud *whit* and a series of descending *keerp* notes; its song is a musical, rolling whistled *whit-will-do*.

Year-round | Adult

SULPHUR-BELLIED FLYCATCHER

Myiodynastes luteiventris L 8½" (22 cm)

FIELD MARKS

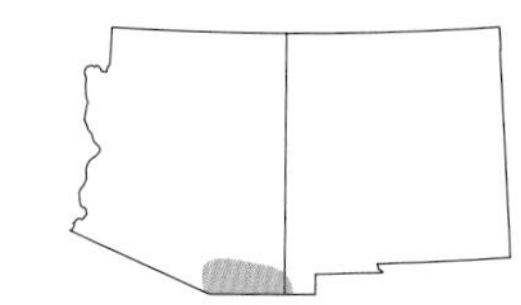

Boldly streaked, above and below; upper parts often show an olive tinge

Rump and tail are rusty red

Underparts pale yellow

Black bill, long and heavy

Behavior

An inconspicuous bird, the Sulphur-bellied Flycatcher often perches quietly high in the canopy, usually alone or in pairs. Feeds on insects and berries. Nesting birds are very territorial, aggressively defending a choice nest cavity. When vocalizing they are conspicuous. Loud call is an excited chatter, sounding something like the squeaking of a rubber duck. Song is a soft *tre-le-re-re.*

Habitat

Woodlands of mountain canyons at elevations from 5,000 to 6,000 feet, particularly where there are sycamore trees. It is a very late arrival in spring—most do not appear until late May or early June.

Local Sites

Uncommon and local summer resident in areas of extreme southeastern Arizona and Guadalupe Canyon in New Mexico. Found in all the major canyons such as Madera, Garden, Miller, and Cave Creek.

FIELD NOTES The Sulphur-bellied is primarily a tropical species whose range just reaches the U.S. It nests throughout the tropical lowlands of Middle America, but all individuals migrate to South America—primarily the Amazon Basin—where they spend the winter months.

Year-round | Adult

WESTERN KINGBIRD

Tyrannus verticalis L 8¾" (22 cm)

FIELD MARKS

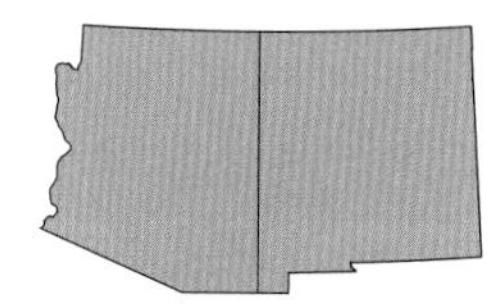

Ashy-gray head, neck, breast

Back tinged with olive

Black tail with white sides

Lemon-yellow belly

Behavior

Feeding almost exclusively on flying insects, leaves its perch to snatch prey in midair, often returning to the perch to eat. Perches horizontally instead of upright. Courtship display involves aerial flight. Like most kingbirds, builds cup-shaped nest near the end of a horizontal tree branch, lining it with weeds, moss, and feathers. Common and gregarious, nesting pairs regularly share the same tree. Call is a single or repeat-edsharp *kip*; also a staccato trill.

Habitat

Common in dry, open country. Has also adapted to more developed areas. Often perches on fence posts and telephone lines.

Local Sites

Very common summer resident and migrant, mainly at lower elevations throughout both Arizona and New Mexico. Sometimes abundant migrant in agricultural areas such as Sulphur Springs Valley.

FIELD NOTES Cassin's Kingbird, *Tyrannus vociferans* (inset), a common summer resident in the montane canyons, has darker gray upperparts than the Western, and a contrasting white chin. Call is a loud *chi-bew*.

Year-round | Adult

LOGGERHEAD SHRIKE

Lanius ludovicianus L 9" (23 cm)

FIELD MARKS

Head and back are bluish-gray; white underparts, very faintly barred

Rump gray or whitish

Seen in flight, wings and tail are dark with white wing patches

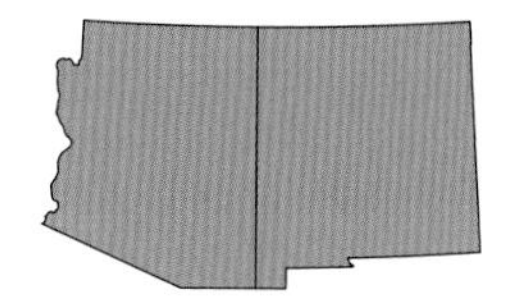

Behavior

Hunts in open or brushy areas, dropping from a low perch, then rising swiftly to the next lookout. Preys on large insects and small vertebrates—mice, lizards, snakes, and small songbirds—killing them with strong, tooth-like bills. With their sharp vision, Loggerhead Shrikes can spot prey 70 yards away and can carry, using their bills, more than their own weight in flight. To compensate for their lack of talons, Loggerhead Shrikes sometimes impale prey on thorny bushes or barbed wire. Song is a medley of low warbles and harsh, squeaky notes. Calls include a harsh *shack, shack.*

Habitat

Open country, lowland plains, grassy pastures, or hill-sides with short grass and low shrubs and trees. Builds cup-like nest in thick shrubs or low trees.

Local Sites

Common along fence lines in agricultural areas such as the Sulphur Springs Valley. Scan any low perches.

FIELD NOTES Loggerhead Shrike populations have been decreasing around the country, but their numbers in the Southwest appear to be stable.

Year-round | Adult

BELL'S VIREO

Vireo bellii L 4¾" (12 cm)

FIELD MARKS

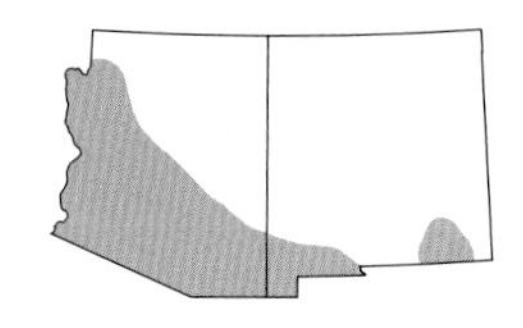

Pale olive-gray back without distinctive markings

Pale wing bars indistinct

Whitish underparts; pale buff-colored sides.

Indistinct white spectacles

Behavior

An energetic bird, hops about low brush looking for insects, and sometimes takes them mid-air. Often flicks its wings while eating. Bell's Vireos have been observed holding a hard-bodied insect with a foot while hammering it with its bill to soften the prey before eating it. The Bell's Vireo is active, and rather secretive. A persistent singer, its song is a series of harsh, scolding notes, sometimes rising at the end, sometimes falling.

Habitat

Moist woodlands, bottomlands, streamside mesquite thickets, desert willows, forest edges. Cup-shaped nest is often hung from fork-shaped branches two to five feet above the ground.

Local Sites

Found at Guadalupe Canyon, along the Gila River, and at Rattlesnake Springs in New Mexico, and along Sonoita Creek and the San Pedro River in Arizona.

FIELD NOTES The slightly larger Warbling Vireo, *Vireo gilvus* (inset), is a widespread migrant and summer resident in aspen groves at higher elevations. It lacks the wingbars of Bell's Vireo and has a prominent white eyebrow.

Year-round | Adult

PLUMBEOUS VIREO

Vireo plumbeus L 5¼" (13 cm)

FIELD MARKS

Prominent white spectacles and white throat; heavy bill

Two white wing bars

Gray upperparts; white below

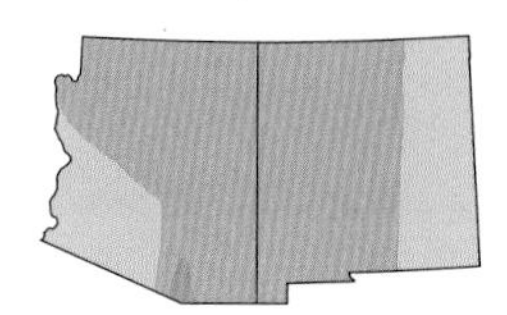

Behavior

Eats mainly insects, which it gleans from twigs and foliage, foraging in a slow, deliberate manner. Sometimes hovers and catches insects on the fly. Will also eat some fruit. Hoarse song of burry two- and three-syllable phrases.

Habitat

Riparian woodlands, montane conifer, and mixed forests. Cup-shaped nest is usually built in a coniferous tree. Both parents incubate three to five eggs for up to two weeks.

Local Sites

Common summer resident in lower montane canyons and pine forests in all the mountain ranges in both states. Fairly common migrant elsewhere, with small numbers regularly wintering at low elevations in the south. Common in canyons such as Madera, Miller, and Cave Creek.

FIELD NOTES The Gray Vireo, *Vireo vicinior* (inset), is a summer resident in pinyon-juniper habitat across much of central Arizona and New Mexico. It is also gray above and white below, but look for its single, thin wing bar and narrow white eye ring.

Year-round | Adult

HUTTON'S VIREO

Vireo huttoni L 5" (13 cm)

FIELD MARKS

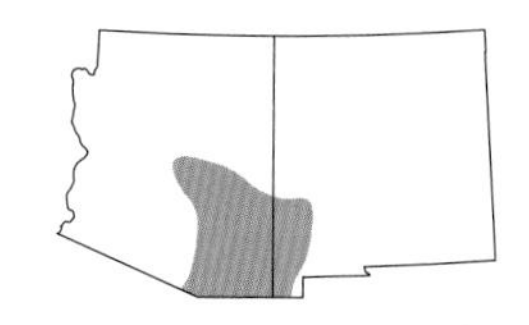

Chunky shape with large head; mostly grayish green plumage

Eye ring blends into pale areas at base of bill

Two white wing bars

Behavior

The Hutton's Vireo moves sluggishly through the canopy foraging for insects among the foliage. Courting males posture before females, fluffing their feathers, fanning their tails and giving a whining or snarling call. Song is a repeated or mixed rising *zuh-wee* and a descending *zoe zo*; also a flat *chew*.

Habitat

Woodlands, primarily evergreen oaks, and in canyons. Nest is a hanging cup bound with spiders and lined with feathers and moss, suspended from a shrub or small tree branch. Female lays three or four white eggs.

Local Sites

Fairly common resident in the mountains of southeastern Arizona and southwestern New Mexico. Look for Hutton's Vireo in mountain canyons such as Madera, Garden, and Cave Creek in Arizona, and in the Peloncillo and Pinos Altos Mountains in New Mexico.

FIELD NOTES The Ruby-crowned Kinglet, *Regulus calendula* (inset), has similar coloring, an eye ring, two white wingbars and the same habit of flicking its wings. Look for the Kinglet's thinner bill and different wing pattern—a prominent dark bar below the lower white wing bar.

Year-round | Adult

STELLER'S JAY

Cyanocitta stelleri L 11½" (29 cm)

FIELD MARKS

Black head, crest, and bill; white feathers above eye and on forehead

Dark gray back, neck, and breast

Purplish blue upperparts; smoky-blue underparts

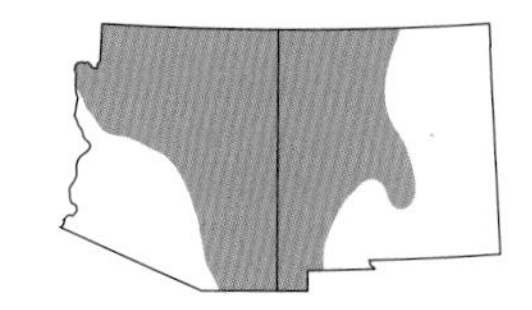

Behavior

Bold and aggressive, the only crested jay in the West. Regularly seen in flocks and family groups, feeding from the treetops to the ground. Often scavenges in campgrounds and picnic areas. Powerful bill efficiently handles a varied diet. Forages during warm months on insects, frogs, carrion, young birds, and eggs. Winter diet is mainly acorns and seeds. Hides food for later consumption. Highly social, jays will stand sentry, ready to mob predators, while others in the flock forage. Calls include a series of *shack* or *shooka* notes.

Habitat

Common in pine-oak woods and coniferous forests, mainly at higher elevations.

Local Sites

Common in the Sandia Mountains in New Mexico, and at both rims of the Grand Canyon, and in the Santa Catalina, Huachuca, and Chiricahua Mountains.

FIELD NOTES The more southern interior west populations have white markings on the forehead and around the eye. In the Pacific Northwest most birds lack these white marks, while in California and the Great Basin, Steller's Jays have blue streaks on their heads.

Year-round | Adult

MEXICAN JAY

Aphelocoma ultramarina L 11½" (29 cm)

FIELD MARKS

Dull blue head without a crest

Upperparts dull blue, slightly grayer on back

Uniform pale gray underparts

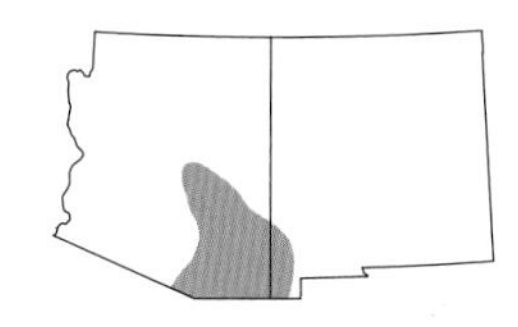

Behavior

Eats mainly acorns and pinyon seeds, but also insects and small vertebrates such as lizards, snakes, and occasionally mice. Call a loud ringing *week,* singly or in a series.

Habitat

Wooded mountain canyons and oak forests. Mexican Jays are communal breeders; their coarse twig nests are built near other Mexican Jay nests, usually in pinyon, pine, oak, or other trees on slopes or mountain tops or in canyons. The entire colony defends nests against predators.

Local Sites

Locally common in pine-oak habitat at middle elevations (up to 7,000 feet) in southeastern Arizona and southwestern New Mexico. In New Mexico, look for it in the Peloncillo and Animas Mountains. Best localities in Arizona include any of the southeast canyons, including Madera, Garden, and Cave Creek.

FIELD NOTES The Western Scrub-Jay, *Aphelocoma californica* (inset), a more widespread species in both Arizona and New Mexico, has similar plumage, but the Mexican Jay is distinguished by the absence of a white eyebrow, lack of dark necklace, and by its chunkier shape.

Year-round | Adult

PINYON JAY

Gymnorhinus cyanocephalus L 10½" (27 cm)

FIELD MARKS

Blue overall; blue throat streaked with white

Long spiky bill and short tail

Lacks the nostril bristles most jays have

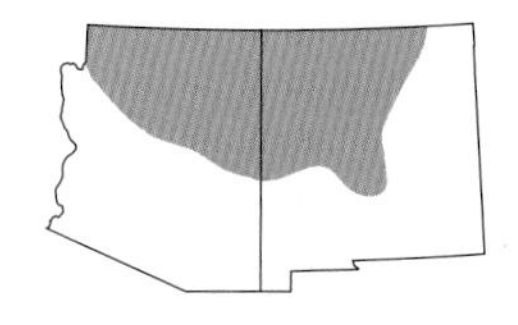

Behavior

Usually seen in large, roaming flocks—almost never singly. Flight is direct, with rapid wingbeats. Diet consists chiefly of pine seeds, which it removes from pine cones opened with its sharp bill. Stores the seeds it does not eat. Also eats some insects and fruits. Typical flight call is a high-pitched, piercing *mew* that can be heard over long distances. Also gives a rolling series of *queh* notes.

Habitat

Pinyon-juniper woodlands of mountains and high plateaus and in ponderosa pine woodlands. Nests in colonies. Nest is cup-shaped and made of grass, bark, and pine needles, placed on a platform of twigs three to six feet high in a pine, oak, or juniper tree.

Local Sites

Most common around Flagstaff and in the White Mountains and in the Black Range and Sacramento Mountains.

FIELD NOTES An irregular wanderer depending on food supply. Sometimes forms large wandering flocks that move into southern Arizona and New Mexico during the fall and winter.

Year-round | Adult

COMMON RAVEN

Corvus corax L 24" (61cm)

FIELD MARKS

Glossy black overall

Large, heavy bill with nasal bristles on top

Shaggy throat feathers

Wedge-shaped tail in flight

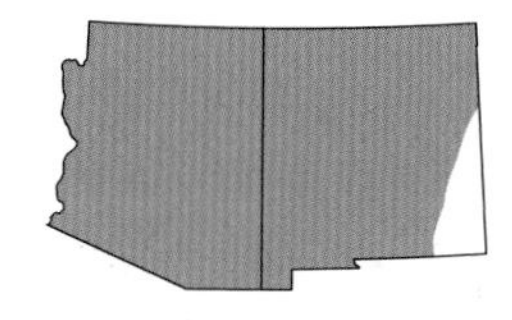

Behavior

The largest perching bird in North America, the Common Raven forages on a great variety of food, from worms and insects to rodents and eggs to carrion and refuse. Small groups are known to hunt together in order to overcome prey that is too large for just one bird to take. Monogamous for life, these birds engage in acrobatic courtship flights of synchronized dives, chases, and tumbles. Calls are extremely variably pitched, from a deep croak to a high, ringing *tok*.

Habitat

The raven can be found in a variety of habitats, but is more abundant at higher elevations. Builds nest high up in trees or on cliffs near water.

Local Sites

Common permanent resident at all elevations and in a variety of habitats throughout much of Arizona and New Mexico. Common on the outskirts of Phoenix and Tucson.

FIELD NOTES The American Crow, *Corvus brachyrhynchos* (inset), found across northern Arizona and New Mexico, is much smaller and lacks the Common Raven's wedge-shaped tail. Readily identified by the familiar *caw* call.

Year-round | Adult male

HORNED LARK

Eremophila alpestris L 6¾-7¾" (17-20 cm)

FIELD MARKS

Pale forehead bordered by black band, ending in hornlike tufts

Black cheek stripes

Pale yellow to white throat and underparts; brown upperparts

Sandy wash on sides and flanks

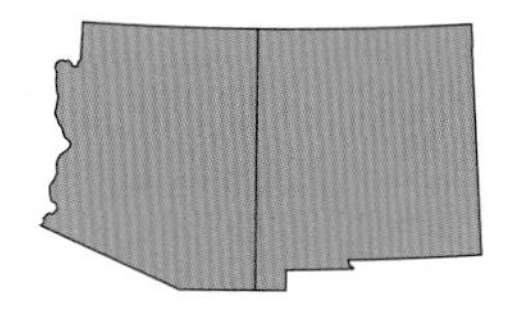

Behavior

Forages on ground, favoring open agricultural fields with sparse vegetation. Feeds mainly on seeds, grain, and some insects. Seldom alights on trees or bushes. On the ground, the Horned Lark walks rather than hops. Song is a weak twittering; calls include a high *tsee-ee* or *tsee-titi*. Females build the nest and incubate the clutch; both sexes feed the young. Outside breeding season, the birds form large flocks that often number in the hundreds.

Habitat

Found in dirt fields, gravel ridges, and grasslands.

Local Sites

Fairly common and widespread summer resident in open grassland habitats across both Arizona and New Mexico. Locally abundant migrant and winter resident in agricultural areas. Look for large wintering flocks in fields in all of the lowland agricultural valleys, such as Sulphur Springs, Rio Grande, and Pecos.

FIELD NOTES The only native lark in North America, the Horned Lark is widespread, with about 20 subspecies (over 40 worldwide). Plumage tends to match the color of the soil it nests on.

Year-round | Adult male

VIOLET-GREEN SWALLOW

Tachycineta thalassina L 5¼" (13 cm)

FIELD MARKS

Dark above with green sheen on head and back; violet sheen on nape, wings, and tail

White below and on face, extending above eye

Pointed wings; notched tail

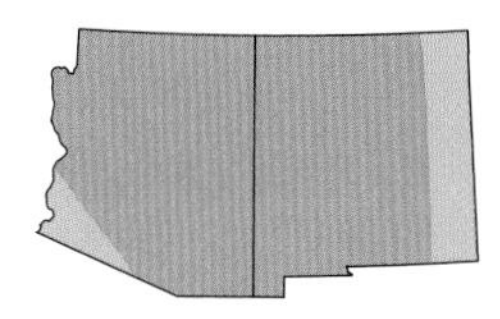

Behavior

Usually feeds in a flock on flying insects by darting close to the ground or low over water, but may also be seen hunting at greater heights. Perches in long rows high up in trees and on fences or wires. Call is a rapid, twittering *chi-chit;* song is a repeated *tsip-tsip-tsip,* most often given in flight around dawn.

Habitat

Found in various woodland settings and adjacent urban areas. Nests in dead trees, abandoned woodpecker holes, rock crevices, and man-made nest boxes. Diligently lines its nest with white feathers.

Local Sites

Common summer resident and migrant, particularly near water. Breeds at higher elevations in mountain ranges such as the Santa Catalina and Chiricahuas.

FIELD NOTES The Tree Swallow, *Tachycineta bicolor* (inset: adult), which is also white underneath, is a more local breeder at high elevations across northern Arizona and New Mexico, and is also a very common migrant throughout. The Tree Swallow's white cheek patch does not extend above its eye, and its back is blue, not green.

Year-round | Adult male

BARN SWALLOW

Hirundo rustica L 6¾" (17 cm)

FIELD MARKS

Long, deeply forked tail

Reddish brown throat

Iridescent blue-black upperparts and breast band

Cinnamon or buff underparts

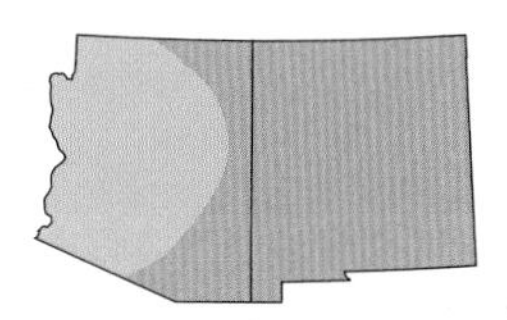

Behavior

An exuberant flyer, it is often seen in small flocks skimming low over the surface of a field or pond, taking insects in midair. Will follow tractors and lawn mowers to feed on flushed insects. Nests in pairs or small colonies.

Habitat

Frequents farms and open fields, especially near water. Has adapted to humans to the extent that it now nests almost exclusively in structures such as barns, bridges, culverts, and garages. Nest is bowl-shaped, made of mud, and lined with grass and feathers.

Local Sites

Common summer resident across the lower elevations of New Mexico and mainly southeastern Arizona. Common migrant throughout. Often nests in towns such as Patagonia and Nogales.

FIELD NOTES The Cliff Swallow, *Petrochelidon pyrrhonota* (inset), also a common summer resident in lowland areas of both Arizona and New Mexico, is distinguished by its squarish tail, buffy rump, and whitish forehead. It nests colonially in gourd-shaped mud nests.

Year-round | Adult

MOUNTAIN CHICKADEE

Poecile gambeli L 5¼" (13 cm)

FIELD MARKS

Unique white eyebrow differs from other chickadees, may be hard to see in summer

Black cap; black bib; white cheeks

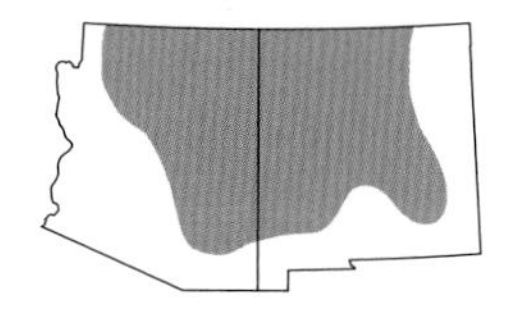

Behavior

The chickadee species most often seen in the higher mountains of the West. In fall and winter forages in loose flocks with other forest songbirds. Seeds and insects are hidden under bark, in pine needle clusters, and in the ground. Call is a hoarse *chick-adee-adee-adee*. Typical song is a three- or four-note descending whistle, *fee-bay-bay* or *fee-bee, fee-bee.*

Habitat

High elevation coniferous and mixed woodlands. Nest is a natural cavity or woodpecker hole; if no cavity is available, the bird digs a hole in decayed wood.

Local Sites

Coniferous forests across most of the higher mountain ranges. In southeastern Arizona, restricted to the Santa Catalina, Rincon, and Pinaleno Mountains.

FIELD NOTES The similar Mexican Chickadee, *Poecile sclateri* (inset), is found in the U.S. only in the Chiricahua Mountains of Arizona and the Animas Mountains of New Mexico. The Mexican Chickadee has a more extensive black bib and lacks the white eyebrow of the Mountain Chickadee.

Year-round | Adult

BRIDLED TITMOUSE

Baeolophus wollweberi L 5¼" (13 cm)

FIELD MARKS

Gray above; whitish below

Distinct crest is gray bordered with black; a gray "bridle" joins the eye line and the throat patch

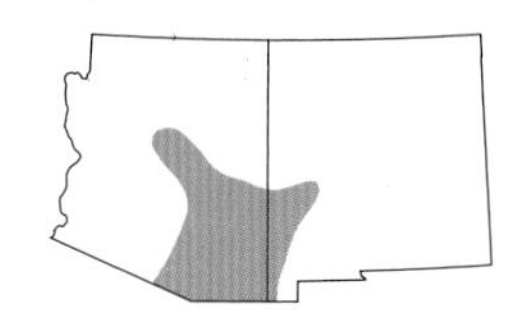

Behavior

Forages in crevices in bark and on tree trunks and branches for insects and their eggs. Nests in cavities, as do all chickadees and titmice, most often in an oak, where it lays five to seven white eggs. Call is a rapid, high-pitched variation of *chick-a-dee-dee.*

Habitat

Stands of oak, pine, juniper, and sycamore at middle elevations. In winter, the birds may move down to streamside cottonwood areas.

Local Sites

Common permanent resident in oak woodland and sycamore riparian areas in southeastern Arizona and southwestern New Mexico. Easily found in any of the montane canyons in southern Arizona, such as Madera, Ramsey, Garden, and Cave Creek.

FIELD NOTES Similar in size, structure, and call, the Juniper Titmouse, *Baeolophus ridgwayi* (inset) is found more in pinyon-juniper habitat.The Juniper Titmouse lacks the distinctive facial pattern of the Bridled Titmouse.

Year-round | Adult

VERDIN

Auriparus flaviceps L 4½" (11 cm)

FIELD MARKS

Yellow face and throat

Gray above; grayish white below

Chestnut-red shoulder patches

Short, black bill

Juvenile brownish gray overall

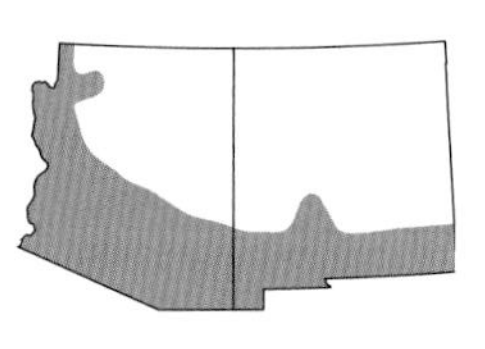

Behavior

Forages for insects, larvae, spiders, berries, and fruit by actively combing branches and foliage, sometimes suspending itself upside-down in chickadee-like fashion. Male builds multiple nests a season, from which female chooses one to use for brooding. Some are reused in successive years, and those not utilized for raising young serve as nighttime roosts or daytime shelters from the blistering southwestern sun. Song is a plaintive three-note whistle, the second note higher-pitched. Repeated *chip* call used to keep small family groups together.

Habitat

Found in mesquite and brushy deserts. Nest is a thick spherical construct up to eight inches wide incorporating sometimes thousands of sticks and twigs.

Local Sites

Lowland desert habitats across southern Arizona and New Mexico. Look for it at the Arizona-Sonoran Desert Museum and Saguaro National Park.

FIELD NOTES The only North American bird of the Remizidae family, the Verdin builds a distinctive ball-like nest. Individuals sometimes construct similar but smaller nests which they use for shelter on winter nights.

Year-round | Adult female

BUSHTIT

Psaltriparus minimus L 4½" (12 cm)

FIELD MARKS

Tiny bird with a noticeably long tail; gray above, paler below

Males have dark eyes; females have pale eyes

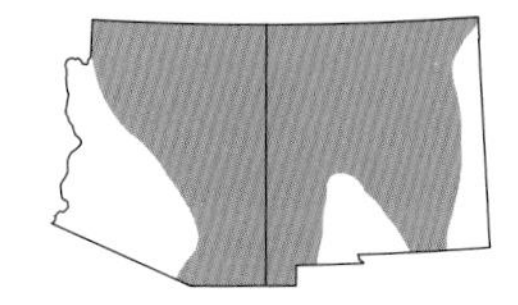

Behavior

A highly sociable bird, the Bushtit often travels in flocks of up to 50 birds, except at nesting time. Flocks move about in a straggling fashion, sometimes in single file, sometimes in small groups. Clings to vegetation, somewhat in the manner of a chickadee as it feeds, often upside down or fluttering briefly, to pluck small insects from the foliage. Its constant calls have a ticking or twittering quality. They weave together a six- to ten-inch-long pendant nest of dried vegetation, moss, lichen, and spider webs.

Habitat

Common breeder in chaparral habitat of oak and juniper.

Local Sites

Fairly common permanent resident in lower montane habitats across much of Arizona and New Mexico. Groups often move into riparian areas, such as along Sonoita Creek and the San Pedro River during the winter.

FIELD NOTES Flocks of foraging Bushtits are always on the alert for predators and will give a loud chorus of rapid "sizzling" notes as an alarm. This usually indicates that a hawk is in the vicinity.

Year-round | Adult male

WHITE-BREASTED NUTHATCH

Sitta carolinensis L 5¾" (15 cm)

FIELD MARKS

Black cap

All-white face and breast

Thin, black bill, tip slightly upturned

Blue-gray upperparts

Rust below to variable extent

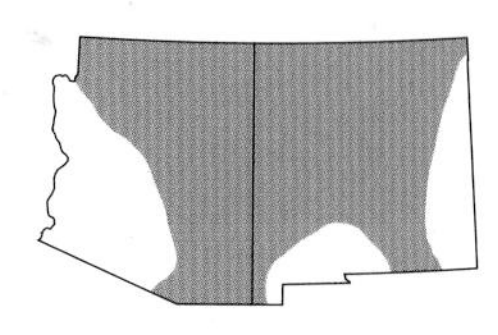

Behavior

An active, nimble feeder. Often spirals around a tree trunk, head down, foraging for insects in bark crevices. Readily visits backyard feeders, preferring sunflower seeds. In this region gives a very short, high-pitched, nasal call with a somewhat laughing quality—very different from eastern or Pacific coast birds. Notes are given in a rapid series, *nyeh-nyeh-nyeh-nyeh.*

Habitat

The White-breasted Nuthatch prefers wooded areas full of oaks and conifers. Nests in woodpecker holes or natural cavities in decaying trees.

Local Sites

Fairly common permanent resident in coniferous forests and montane riparian areas throughout the mountains of both states. Common in canyons such as Madera, Garden, and Cave Creek.

FIELD NOTES The White-breasted is the most widely distributed nuthatch in North America, and call varies regionally. Eastern birds give a slow, low-pitched *yank;* Pacific birds are somewhat similar but higher pitched, longer, and harsher *eerh, eerh;* interior west birds are described above.

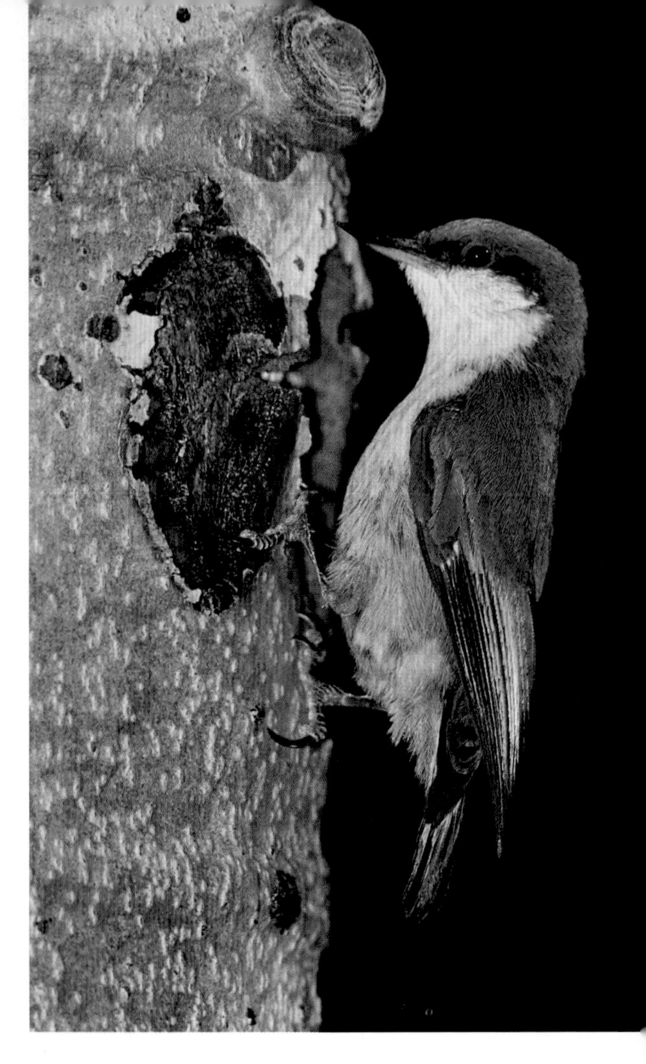

Year-round | Adult

PYGMY NUTHATCH

Sitta pygmaea L 4¼" (11 cm)

FIELD MARKS

Tiny; short-tailed

Blue-gray back; white to buff underparts

Gray-brown cap extends down to eyes

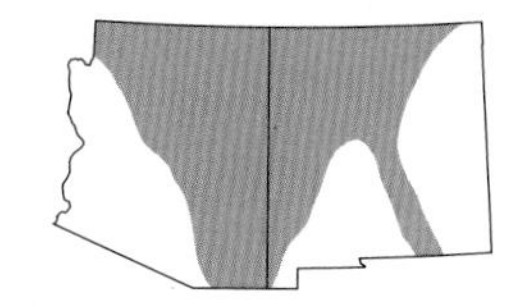

Behavior

Hops along tree trunks and branches and forages in clusters of pine needles at the tips of branches and in bark crevices; sometimes flies to the ground to feed. Hides and stores food. Roams in loose flocks. Typical calls include a high, rapid *peep-peep* and a piping *wee-bee.* Smaller than other nuthatches.

Habitat

Yellow and ponderosa pine forests. Both sexes excavate a small cavity, usually near the top of a dead pine or upright post. Female lays up to eight speckled reddish brown eggs, which hatch in about 16 days. Offspring from previous years help their parents raise the young.

Local Sites

Common in pine forests across the mountains of both Arizona and New Mexico. Very common at the Grand Canyon, high up in the Santa Catalina and Chiricahua Mountains in Arizona, and south to the Sacramento Mountains in New Mexico.

FIELD NOTES The high-note call of the Pygmy Nuthatch contrasts with the distinctive nasal, tin-horn call of the Red-breasted Nuthatch, *Sitta canadensis* (inset: female top, male bottom), which has rust underparts and a white eyebrow.

Year-round | Adults

CACTUS WREN

Campylorhynchus brunneicapillus L 8½" (22 cm)

FIELD MARKS

Dark crown; broad, white eyebrow

Streaked back; heavily barred wings and tail

Breast densely spotted with black

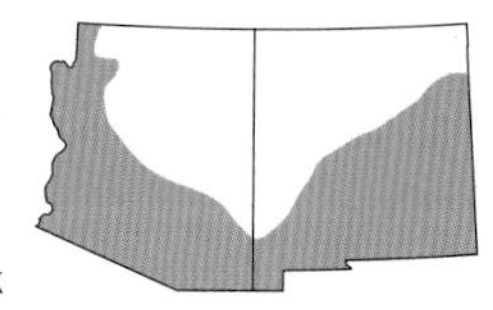

Behavior

Often seen in pairs or small family groups, foraging for food on the ground or gleaning insects from vegetation. Its song, heard year-round, is a harsh, low-pitched, and rapid *cha-cha-cha-cha-cha.* During courtship, a male wren may chase a visiting female as though she were an intruder, but the female's call of *chur* announces that she is interested in him. The male may then hop stiffly around her and fly to or sing around one of his nest sites, inviting her to inspect it.

Habitat

Common in cactus country and arid hillsides and valleys.

Local Sites

Common permanent resident in cactus-dominated desert across southern Arizona and New Mexico. Very common at Saguaro National Park, and widespread wherever there are stands of cholla cactus, including the Rio Grande Valley and eastern plains of New Mexico.

FIELD NOTES North America's largest wren, the Cactus Wren is part of a widespread genus that has species extending into South America. It builds its domed nest in cholla cactus or other thorny plants, whose spines protect the eggs and young from predators. After the young have fledged, the nests are maintained as roosting sites.

Year-round | Adult

CANYON WREN

Catherpes mexicanus L 5¾" (15 cm)

FIELD MARKS

Flat head and long bill; white throat and breast

Chestnut tinged body with fine speckles

Long, chestnut tail with thin black bars

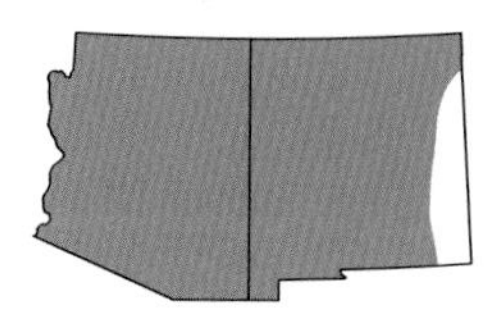

Behavior

Extracts insects from crevices in rock with long bill; occasionally attempts flycatching. Can climb up, down, and across rocks. Pairs remain together throughout the year. Loud, silvery song is a decelerating, descending series of liquid *tees* and *tews*. Typical call is a sharp *jeet*.

Habitat

Canyons, boulder piles, and cliffs, often near water. Cup-shaped nest of twigs and grasses is built by both male and female. Female incubates five white eggs for 12 to 18 days, while the male feeds her.

Local Sites

Fairly common permanent resident near cliffs and rocky slopes in the mountains and canyons across both states. Look for it on the rocky cliffs above the Patagonia Roadside Rest and along the Ruby Road near Nogales.

FIELD NOTES The Rock Wren, *Salpinctes obsoletus* (inset), found in similar habitats, lacks the reddish-brown tones and contrasting white throat of the Canyon Wren and has a shorter bill.

Year-round | Adult

BEWICK'S WREN

Thryomanes bewickii L 5¼" (13 cm)

FIELD MARKS

Gray-brown upperparts

Whitish underparts

Long, white eyebrow; long decurved bill

Long rounded tail barred lightly in black, edged in white

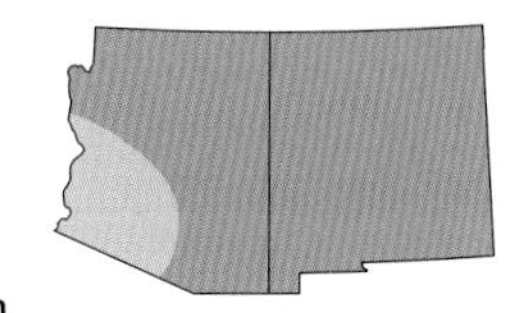

Behavior

The Bewick's Wren is often seen in pairs. Feeds primarily on ground but also gleans insects and spiders from vegetation. Holds tail high over back, flicking it often from side to side. Its song is a variable, high, thin, rising buzz, followed by a slow trill. Calls include a flat, buzzy *jip*.

Habitat

Prefers brushland and open woods. Inquisitive and tame, the Bewick's Wren can usually be found around human habitation such as ranches and farms. Nests in a variety of cavities from hollow logs to mailboxes.

Local Sites

Common, widespread, permanent resident in variety of habitats from desert washes to lower montane canyons across much of the region. Look for it along Sonoita Creek and the San Pedro River and the Rio Grande.

FIELD NOTES The House Wren, *Troglodytes aedon* (inset), which breeds in higher mountains throughout Arizona and New Mexico and is a common migrant and winter visitor to riparian areas, where it overlaps with Bewick's, has a shorter tail and lacks the white eyebrow of the Bewick's Wren.

Breeding | Adult male

BLACK-TAILED GNATCATCHER

Polioptila melanura L 4" (10 cm)

FIELD MARKS

Blue-gray above, whitish below

White terminal spots on graduated, black tail feathers

Male has black cap

Short bill

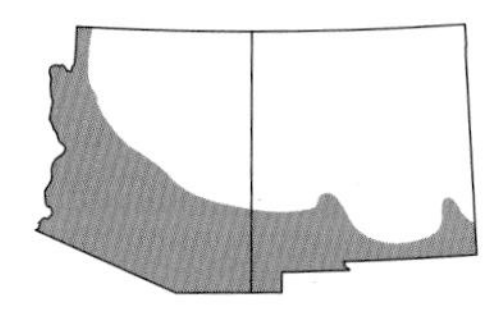

Behavior

Hops and flits among foliage, flipping its long trail as it forages for tiny insects. Lives in pairs at all seasons. Calls include rasping *chee* and hissing *ssheh.* Song is a rapid series of *jee* notes.

Habitat

Desert; partial to washes, creosote bush flats, mesquite thickets. Both sexes build an open-cup nest in a low shrub, using weeds, grass, strips of bark, spider webs, and plant fibers. Both parents incubate three to five bluish white eggs with red-brown dots for 14 days.

Local Sites

Fairly common permanent resident in lowland desert habitats across southern Arizona and extreme southern New Mexico. Look for it in desert areas around Phoenix and Tucson, particularly at the Arizona-Sonoran Desert Museum and at the San Simone Cienaga in New Mexico.

FIELD NOTES The Blue-gray Gnatcatcher, *Polioptila caerulea* (inset: female top, male bottom), breeds at higher elevations but winters in the same areas as the Black-tailed Gnatcatcher. The Blue-gray has a longer bill and more extensive white on the undersurface of its tail.

Year-round | Adult male

WESTERN BLUEBIRD

Sialia mexicana L 7" (18 cm)

FIELD MARKS

Chestnut shoulders, upper back

Deep purple-blue upperparts and throat in male; duller, brownish gray in female

Chestnut breast, sides, and flanks in male; chestnut gray in female

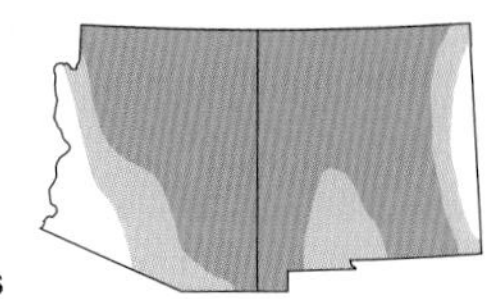

Behavior

Using large, acute eyes, a bluebird will hunt from a perch high above the ground, swoop down to seize crickets, grasshoppers, and spiders, which it may have spotted from as far away as 130 feet. The call note of the Western Bluebird is a mellow *few,* extended in brief song to *few few fawee.*

Habitat

Found in woodlands, farmlands, orchards, and deserts in the winter. Nests in holes in trees or posts as well as in nest boxes. Frequents mesquite-mistletoe groves in winter.

Local Sites

Most commonly seen in the Grand Canyon, across the Mogollon Rim in central Arizona, and in all of the mountains of New Mexico south to the Guadalupe Mountains.

FIELD NOTES The Mountain Bluebird, *Sialla currucoides* (inset: female left, male right), is found in open grasslands at higher elevations in the mountains of Arizona and New Mexico. The male lacks any rust color and differs from the Western in its sky blue breast color. Females and immatures are grayer with only a hint of rust.

Year-round | Adults

AMERICAN ROBIN

Turdus migratorius L 10" (25 cm)

FIELD MARKS

Gray-brown above; darker head and tail

Yellow bill

Brick-red underparts, paler in female

White lower belly

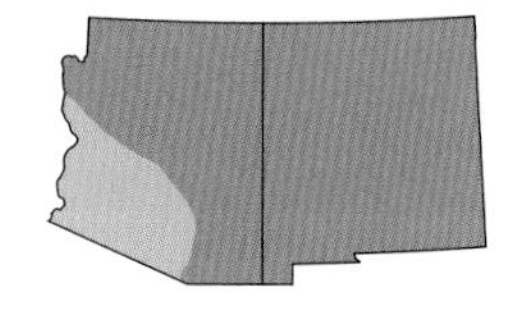

Behavior

Best known and largest of the thrushes. Very often seen on suburban lawns, hopping about and cocking its head to one side in search of earthworms. The American Robin will also glean butterflies, damselflies, and other flying insects from foliage. Even takes prey in flight. Robins also consume fruit, usually in the fall and winter, making them one of the most abundant—and wide-ranging—thrushes.

Habitat

Found in riparian areas in montane canyons, and in coniferous forests across much of Arizona and New Mexico. Winters in lowland riparian areas, orchards.

Local Sites

Common summer resident in montane habitats across most of the mountain ranges in Arizona and New Mexico. More widespread as a migrant and winter resident, when it is found in a variety of habitats in both states.

FIELD NOTES The American Robin is a facultative migrant that is common some years as a winter visitor yet almost absent others, depending on food supply.

Year-round | Adult

HERMIT THRUSH

Catharus guttatus L 6¾" (17cm)

FIELD MARKS

Gray-brown upperparts; white to buffy underparts with dense spotting mostly on breast

Reddish tail contrasts with upperparts

Whitish eye ring; dark lateral throat streak

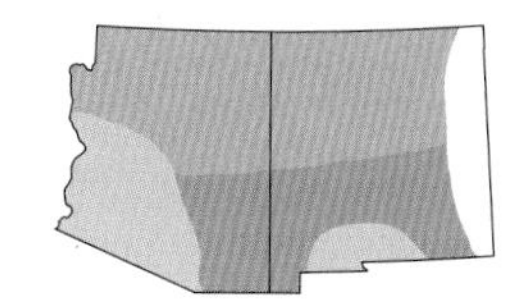

Behavior

The Hermit Thrush is a shy, terrestrial bird that forages on the forest floor for insects or ascends into bushes in search of berries. When interrupted it flies up into a low bush, flicking its wings nervously and slowly raising and lowering its tail. Common call is a blackbird-like *chuck,* often doubled; song is a serene series of clear, flutelike notes, the phrases repeated at different pitches, lending it a lyrical quality.

Habitat

For breeding, coniferous forests typically in areas of relatively little undergrowth. In winter and migration uses a wide variety of habitats, but always with forest or brushy cover nearby, especially the shady portions of major river valleys.

Local Sites

Fairly common summer resident. Common migrant and winter visitor. Check along the San Pedro River in Arizona and the Rio Grande in New Mexico.

FIELD NOTES There is an influx of Hermit Thrushes into southern Arizona and New Mexico from breeding grounds as far north as Alaska. The western subspecies has gray flanks, while eastern birds have warm brown flanks.

Year-round | Adult

NORTHERN MOCKINGBIRD

Mimus polyglottos L 10" (25 cm)

FIELD MARK

Gray plumage; dark wings and tail

White wing patches and outer tail feathers, which flash conspicuously in flight

Repeats same phrase several times while singing

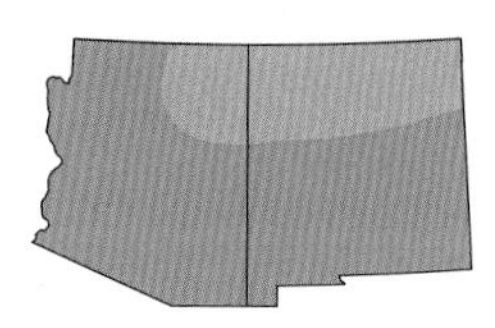

Behavior

A widespread mimic, the Northern Mockingbird learns as many as 200 songs and calls of other species. Males have a spring and fall repertoire. Highly pugnacious, will protect its territory against not only other birds but also dogs, cats, and even humans. A varied diet includes grasshoppers, spiders, snails, earthworms, and berries.

Habitat

Resides in a variety of habitats, including towns. Feeds close to the ground, often in thickets or in heavy vegetation.

Local Sites

Common summer resident in a variety of habitats across much of Arizona and New Mexico. Found commonly in residential areas of most towns and cities, including Phoenix, Tucson, and the cities along the Rio Grande in New Mexico. Winters mainly in the southern half of both states.

FIELD NOTES The unrelated Loggerhead Shrike, *Lanius ludovicianus* (inset), looks strikingly similar to the Northern Mockingbird. In flight, however, the shrike's wings and tail are darker and the white wing patches are smaller.

Year-round | Adult

CURVE-BILLED THRASHER

Toxostoma curviostre L 11" (28 cm)

FIELD MARKS

Gray-brown above, paler below; faint spots on breast

Long, dark, strongly curved bill

Long, dark tail has white tips; orange to yellow-orange eyes

Juvenile bill straighter, eyes paler

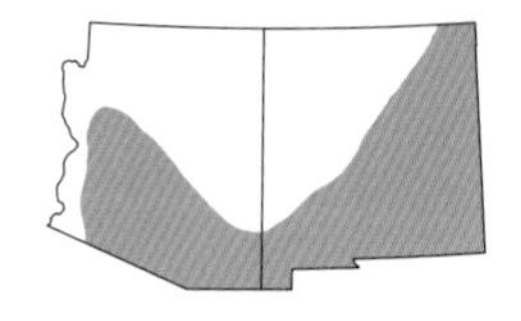

Behavior

The most common desert thrasher; forages by digging into dirt or sand to expose insects, spiders, small reptiles. Also eats cactus fruit, seeds, berries, snails. To obtain necessary moisture in arid desert surroundings, the Curve-billed will visit birdbaths, dripping outdoor faucets, or any other available source of water. Claims territory with a distinctive, sharp *whit-wheet-whit.* Song varies; includes low trills and warbles, often repeated.

Habitat

Canyons, semiarid brushlands, deserts. Builds cup-like nest of twigs and roots in cholla and other dense, thorny vegetation.

Local Sites

Common permanent resident across much of southern Arizona and New Mexico, even entering suburban areas. Most common in the Sonoran Desert regions outside Phoenix and Tucson, particularly Saguaro National Park.

FIELD NOTES The Bendire's Thrasher, *Toxostoma bendirei* (inset), is similar to Curve-billed but has a shorter bill with a pale base that is only slightly decurved, and small triangular spots on its breast.

Year-round | Adult

CRISSAL THRASHER

Toxostoma crissale L 11½" (29 cm)

FIELD MARKS

Bill long, slender, strongly decurved; dark malar streak at base of bill

Gray-brown plumage with chestnut undertail coverts

Long dark tail

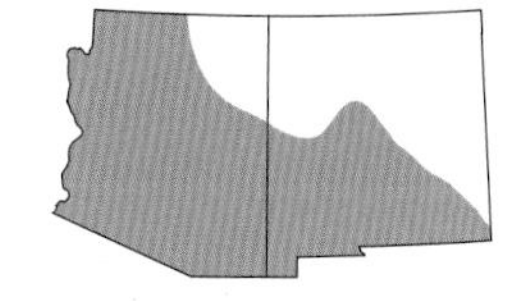

Behavior

Uses its long curved bill to probe for insects and grubs in the ground. Also eats berries and wild grapes. Calls include a repeated *chideery* and a whistled *toit-toit-toit.* Song is a long series of *churrs* and whistles, often mimicking other species.

Habitat

Hides in underbrush. Found mainly in dense mesquite and willows along streams and washes; sometimes on lower mountain slopes. Builds a large, twiggy nest on a branch or in a fork of mesquite trees, also in willow, sagebrush, or other desert shrubs.

Local Sites

Uncommon yet widespread permanent resident in desert washes and lower foothills across much of the region. Look for it in the foothills of the Sandia Mountains above Albuquerque and the Tularos Valley in New Mexico, and in desert scrub habitat in the Sulphur Springs Valley and below Portal in Arizona.

FIELD NOTES The Crissal Thrasher can be very secretive, except in late winter and early spring when the birds sit at the tops of exposed dead limbs and sing.

Summer | Adult

EUROPEAN STARLING

Sturnus vulgaris L 8½" (22 cm)

FIELD MARKS

Iridescent black plumage

White feather tips on head and breast wear off by late spring

Yellow bill, dark in winter

Juvenile dull brown

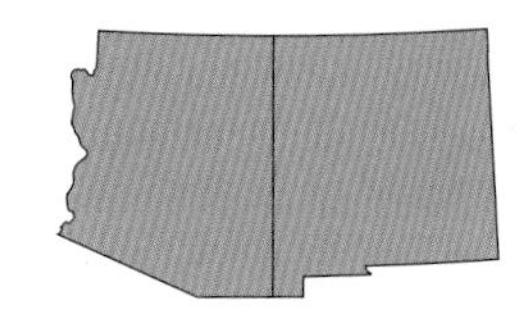

Behavior

A highly social and aggressive bird, the European Starling will gorge on a tremendous variety of food, ranging from invertebrates—such as snails, worms, and spiders—to fruits, berries, grains, seeds, and even garbage. Short, square tail and triangular wings are particularly distinguishable during flight. Will imitate the songs of other species and has call notes that include squeaks, warbles, chirps, and twittering. Outside nesting season, usually seen in large flocks.

Habitat

Adaptable, starlings thrive in a variety of habitats, from urban centers to agricultural regions. They nest in cavities, ranging from crevices in urban settings to woodpecker holes and nest boxes.

Local Sites

Common to abundant permanent resident in cities and towns throughout Arizona and New Mexico. Forms large flocks during the winter.

FIELD NOTES A Eurasian species introduced to North America in the 1890s, starlings have spread throughout the United States and Canada. Starlings often compete for and take over nest sites of other hole-nesting, native species.

Nonbreeding | Adult

AMERICAN PIPIT

Anthus rubescens L 6½" (17 cm)

FIELD MARKS

Breeding birds grayish above, faintly streaked below

White outer tail feathers, visible in flight

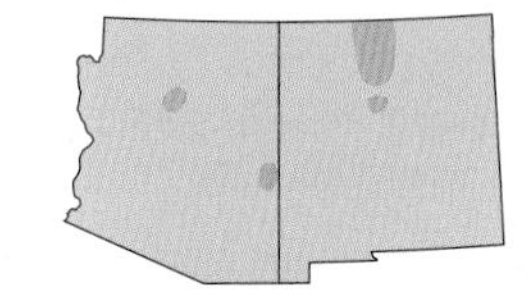

Behavior

When foraging, walks briskly with a strong gait and bobbing head in search of seeds, insects, and arthropods. Also eats emerging insect larvae and takes aquatic insects from shallow water. Sometimes flycatches. Call, given in flight, is a sharp *pip-pit.* Song is a rapid series of *chee* or *cheedle* notes.

Habitat

Farm fields, short-grass meadows, and river and lake margins. Builds cup-shaped nest on the ground. Female incubates the eggs, and is fed by the male during incubation.

Local Sites

Fairly common summer resident above the treeline in the San Francisco Peaks and White Mountains in Arizona and in the Sangre de Cristo Mountains of New Mexico. Locally abundant migrant and winter resident in agricultural areas and grasslands throughout the southern portions of both states.

FIELD NOTES The Rocky Mountain breeding subspecies, which is pale buffy below, nests in Arizona and New Mexico. Numbers are augmented in winter by migrants from the Arctic tundra nesting subspecies. The American Pipit often forms large single-species flocks during the winter.

Year-round | Adult

CEDAR WAXWING

Bombycilla cedrorum L 7¼" (18 cm)

FIELD MARKS

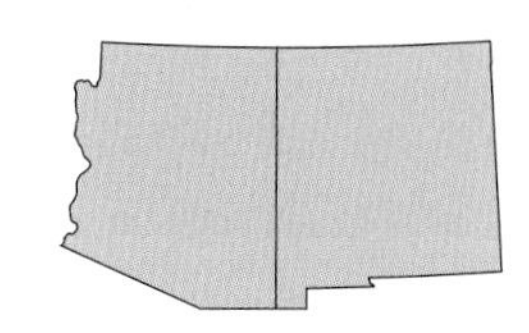

Distinctive, sleek crest

Black mask bordered in white

Silky plumage with brownish chest and upperparts

Yellow terminal tail band

May have red, waxy tips on wings

Behavior

Eats the most fruit of any bird in North America. Up to 84 percent of its diet includes cedar, peppertree, and hawthorn berries and crabapple fruit. Also consumes sap, flower petals, and insects. Cedar Waxwings are gregarious in nature and band together for foraging and protection. Flocks containing from a few to thousands of birds may feed side by side in winter. Flocks rapidly disperse, startling potential predators. Call is a soft, high-pitched trilled whistle.

Habitat

Found in open habitats where berries are available. The abundance and location of berries influence the Cedar Waxwing's migration patterns; it will move long distances only when its food sources run out.

Local Sites

Locally common yet irregular migrant and winter resident across Arizona and New Mexico.

FIELD NOTES Facultative migrants, Cedar Waxwings show up some years and are absent others, depending on food supply locally and farther north. One of its favorite foods is the pyrocantha berry.

Year-round | Adult female

PHAINOPEPLA

Phainopepla nitens L 7¾" (20 cm)

FIELD MARKS

Male shiny black; white wing patch conspicuous in flight

Female is gray

Both sexes have distinct crest; long tail; red eyes

Behavior

Feeds chiefly on insects and mistletoe berries. Defends patches of mistletoe from other birds. Flight is fluttery but direct, often very high. Courtship involves exchanges of berries, as well as special flights and chases. Distinctive call is a querulous, low-pitched, whistled *wurp?* Seldom-heard song is a brief warble.

Habitat

Chapparal, desert washes, pinyon-juniper forests—anywhere mistletoe berries are abundant. Phainopeplas nest in early spring in mesquite brushlands, later moving to cooler, wetter habitats to raise a second brood.

Local Sites

Look for it in desert washes near Phoenix and Tucson, along Sonoita Creek near Patagonia, and at Rattlesnake Springs.

FIELD NOTES A mistletoe specialist, the Phainopepla moves around during migration and winter in search of mistletoe berries. In return, the seeds of the parasitic mistletoe are spread far and wide after they pass through the Phainopepla's digestive tract.

Year-round | Adult male

OLIVE WARBLER

Peucedramus taeniatus L 5¼" (13 cm)

FIELD MARKS

Head, throat, and nape tawny orange-brown; dark mask

Two broad white wing bars; white outer tail feathers

Female has olive crown; yellow face and throat

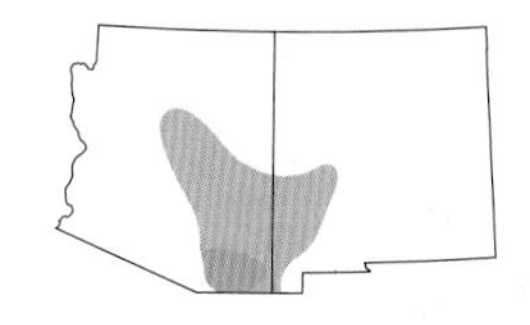

Behavior

Usually found high in conifers, where it creeps slowly and deliberately along branches as it forages for small insects. Kinglet-like wing flickering is distinctive. Typical song is a loud *peeta-peeta-peeta.* Call is a soft, whistled *phew.*

Habitat

Favors open coniferous forests above 7,000 feet. Nests high in pines and firs.

Local Sites

Fairly common summer resident in pine forests of mountain ranges south of the Mogollon Rim in Arizona and in southwest New Mexico. In New Mexico, look for it on Signal Peak in the Pinos Altos Mountains and Emory Pass in the Black Range. In Arizona, it is common on Mount Lemmon near Tucson and near Rustler Peak in the Chiricahua Mountains. Some individuals in the southern portion of the range are permanent residents.

FIELD NOTES Long thought to be an aberrant wood-warbler, recent genetic studies have determined that the Olive Warbler is only distantly related. The Olive Warbler is now placed in its own family with uncertain relationships, possibly to finches or Old World warblers.

Year-round | Adult

ORANGE-CROWNED WARBLER

Vermivora celata L 5" (13 cm)

FIELD MARKS

Olive-green above; pale yellow-green below; brightest yellow on undertail coverts

Dark eyeline; faint yellow eyebrow

Orange crown patch is absent on some females and indiscernible on some males

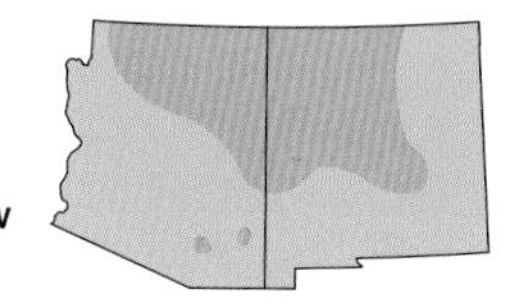

Behavior

Often seen by itself, deliberately foraging low in trees, shrubs, and grasses for insects. Known to respond readily to pishing by birders. Song is a high-pitched staccato trill that slows and drops pitch slightly at the end. Call note is a sharp *chip*. Listen as well for this bird's flight call, a high, clear *seet.*

Habitat

Inhabits open, brushy woodlands, forest edges, and thickets. Sometimes found in suburban areas. Nests on the ground among shrubs or grasses.

Local Sites

Uncommon to common summer resident at high elevations across much of northern Arizona and New Mexico, locally south to the Santa Catalina and Pinaleno Mountains. Common migrant in a variety of habitats throughout both states, and fairly common winter resident in riparian areas in southern Arizona, such as the Santa Cruz River.

FIELD NOTES Different subspecies of these birds winter in Arizona and New Mexico—the gray-headed form that breeds in the Rocky Mountains and the yellow-green form that breeds along the Pacific coast north into Alaska.

Year-round | Adult

LUCY'S WARBLER

Vermivora luciae L 4¼" (11 cm)

FIELD MARKS

Pale gray above, whitish below; short-tailed

Diffuse white eye ring

Male has reddish crown, patch and rump; duller in female

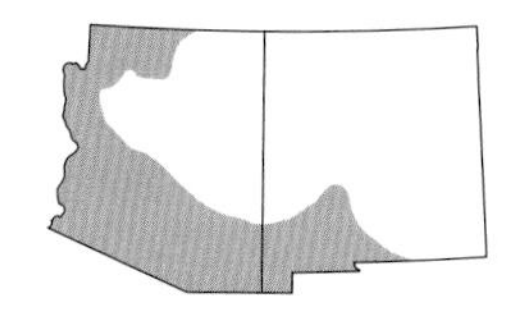

Behavior

This tiny, desert-dwelling wood-warbler often flicks its tail while foraging. Dines mainly on caterpillars, beetles, and leafhoppers. Lively song is a short trill followed by lower, whistled notes. Call is a sharp *chink.*

Habitat

Mesquite and cottonwood along watercourses; shrubby desert areas and desert foothills. Female builds an open-cup nest in an existing tree cavity in mesquite thickets, usually near water, lays four or five speckled white eggs, and incubates them for 12 days.

Local Sites

Common summer resident in riparian areas and adjacent mesquite washes across much of southern Arizona and southwestern New Mexico. Common in desert areas near Phoenix and Tucson, and birding areas along Sonoita Creek and the San Pedro River. Locally common along the Rio Grande Valley.

FIELD NOTES The Virginia's Warbler, *Vermivora virginiae* (inset), is slightly larger than the Lucy's and breeds at higher elevations on oak hillsides. It has a white eye ring, yellow patch on its breast, and yellow undertail coverts.

Year-round | Adult male

YELLOW WARBLER

Dendroica petechia L 5" (13 cm)

FIELD MARKS

Bright yellow overall

Plump and short-tailed

Dark eye prominent in yellow face

Male shows distinct reddish streaks below; streaks faint or absent in female

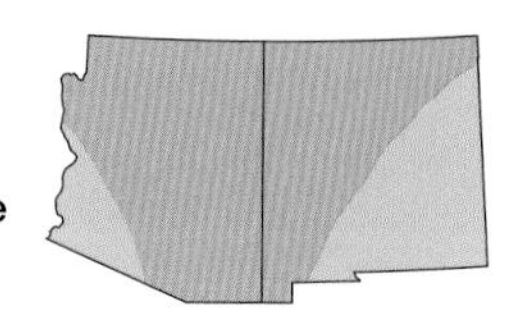

Behavior

The Yellow Warbler frequently pumps its tail and responds vigorously to pishing. Forages in trees, shrubs, and bushes, gleaning insects and larvae from branches and leaves. Will sometimes spot flying insects from a perch and chase them down. Nests in the forks of trees or bushes at eye level or a little higher. Male and female both feed nestlings. Song is a rapid, variable *sweet sweet I'm so sweet.* Call is an emphatic downslurred *tchip.*

Habitat

For breeding they seek out riparian woodlands of the lowlands and foothill canyons, but are also found at higher elevations along watercourses.

Local Sites

Locally common summer resident in riparian areas throughout much of Arizona and New Mexico. A common migrant along rivers such as the San Pedro in Arizona and the Rio Grande in New Mexico.

FIELD NOTES Wilson's Warbler, *Wilsonia pusilla* (inset: male left, female right), is a very common migrant throughout Arizona and New Mexico. Smaller than the Yellow, the male is distinguished by its black cap; both sexes lack bright yellow feather edges on their wings.

Breeding | Adult male "Audubon's"

YELLOW-RUMPED WARBLER

Dendroica coronata L 5½" (14 cm)

FIELD MARKS

Bluish gray head; yellow crown patch and throat; white eye crescents

Yellow rump and flank patch; breeding male has black breast

Females and winter males browner and streaked below

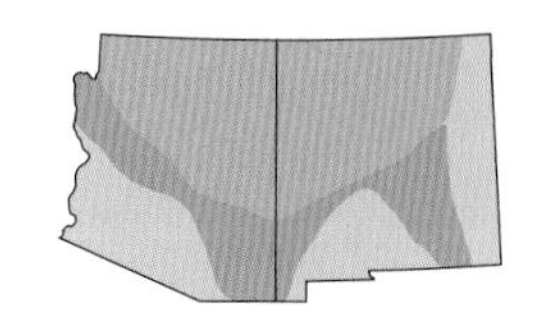

Behavior

The western form of the Yellow-rumped that breeds in our region is known as the "Audubon's Warbler" (see Field Notes). Darts from tree to tree among the branches, foraging for insects and berries; sallies from treetops to catch insects on the fly. Song is a variable slow warble that rises or falls toward the end. Call is a loud, oft-repeated *chip* or *chep*.

Habitat

Coniferous or mixed woodlands. Female lays four or five spotted white eggs in a bulky nest of twigs, roots, and grass lined with hair and feathers and built in a conifer.

Local Sites

Common summer resident at higher elevation in all of the major mountain ranges across both states. Locally abundant migrant and winter resident in riparian areas in the south. Large numbers winter along the Salt River near Phoenix and the Santa Cruz River near Tucson.

FIELD NOTES The eastern form of the Yellow-rumped Warbler, "Myrtle Warbler" (inset), is rare in our region. It has a white throat and pale supercillium rather than the yellowish throat and eye crescents of the "Audubon's."

Year-round | Adult male

BLACK-THROATED GRAY WARBLER

Dendroica nigrescens L 5" (13 cm)

FIELD MARKS

Black and white head

Small yellow spot in front of eye

Gray back streaked with black; sides streaked with black

White underparts and undertail coverts

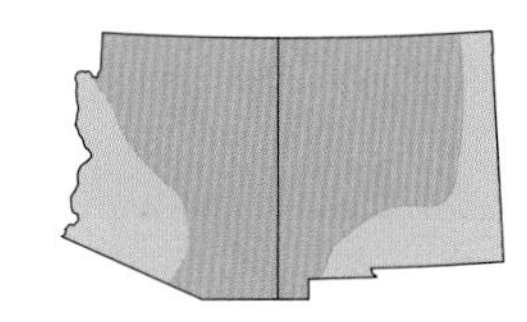

Behavior

Gleans insects, larvae, and caterpillars from low- to mid-level branches, twigs, and leaves. Will join mixed-species foraging flocks in the winter. Varied songs of the Black-throated Warbler include a buzzy *weezy weezy weezy weezy-weet,* with the ultimate or penultimate note higher.

Habitat

Inhabits woodlands, brushlands, and chaparral; mostly coniferous in the north, oaks farther south. Nests alone in forks or branches of trees, fairly close to the ground.

Local Sites

Common summer resident in foothills and lower montane habitats throughout most of the mountain ranges in Arizona and New Mexico. Look for it on the oak hillsides in many of the canyons in southeastern Arizona, such as Madera, Garden, and Cave Creek.

FIELD NOTES Townsend's Warbler, *Dendroica townsendi* (inset: male), a migrant found mainly in the region's higher elevation coniferous forests, is similar in pattern to the Black-throated Gray Warbler, but much of that species' white feathering is replaced with yellow.

Year-round | Adult

RED-FACED WARBLER

Cardellina rubrifrons L 5½" (14 cm)

FIELD MARKS

Distinctive red and black face pattern

Gray back and tail; white rump and underparts

White nape patch

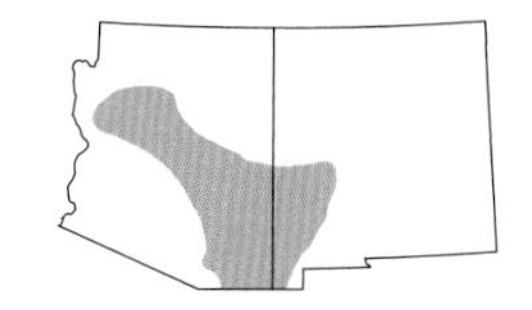

Behavior

An active and energetic bird, the Red-faced Warbler might resemble a chickadee were it not for its bright red face. Keeps to the outside canopy of tall trees, gleaning insects. Spirals downward to catch flying insects; occasionally hovers to pick insects from foliage. Eats caterpillars, flies, and bees. It has a characteristic habit of flicking its tail sideways. Male establishes and defends territory, chasing out intruders. Female defends the territory against other females. Song is a series of varied, ringing *zweet* notes, similar to that of the Yellow Warbler. Call is a loud *chup.*

Habitat

Mountains above 5,000 feet, among fir, spruce, and oaks. The nest is an open cup placed on the ground in a well-concealed depression at the base of a tree or rock.

Local Sites

Found at Bear Wallow on Mount Lemmon and Rustler Park in the Chiricahua Mountains. In New Mexico, look for it at Emory Pass in the Black Range, and in the Pinos Altos Mountains.

FIELD NOTES Winters in the mountains of Mexico, usually in a cloud-forest habitat. During the past 50 years, has expanded its U.S. breeding range to the north and west but still only breeds in Arizona and western New Mexico.

Year-round | Adult

PAINTED REDSTART

Myioborus pictus L 5¾" (15 cm)

FIELD MARKS

Bright red lower breast and belly

Black head and upperparts; bold white wing patch

White outer tail feathers conspicuous when fanned

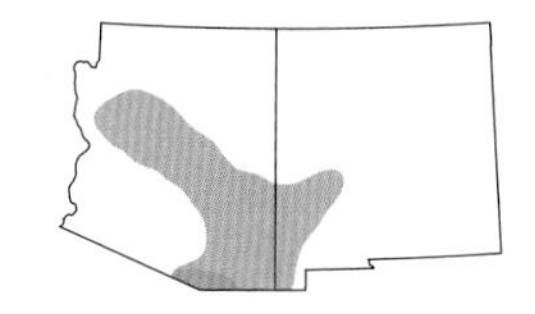

Behavior

The Painted Redstart gleans insects from leaves and tree branches, and also catches them on the wing. While foraging, the bird flashes its white wing patches and fans its tail feathers. This action seems to flush out insects. A courting pair sings to each other, sometimes for hours, at about 5,000 ft to 7,000 feet. Song is a series of rich, liquid warbles. Males sing to defend territory and to attract a mate. Call is a clear, whistled *chee.*

Habitat

Riparian areas and woodlands, particularly in the mountains. Generally nests on the ground; the nest is a well-hidden bulky cup often situated near water.

Local Sites

Common summer resident in all of the southeastern Arizona canyons such as Madera, Garden, and Cave Creek. In New Mexico, most numerous in the Pinos Altos and Animas Mountains.

FIELD NOTES This beautiful warbler is not closely related to the American Redstart, *Setophaga ruticilla*. Small numbers of Painted Redstarts winter in the canyons of extreme southern Arizona, but the vast majority leave the region to spend the winter months in Mexico and Central America.

Year-round | Adult male

SUMMER TANAGER

Piranga rubra L 7¾" (20 cm)

FIELD MARKS

Adult male is rosy red overall

Most females have olive green upperparts, yellow underparts

Some females have overall orangish wash

Large yellowish bill; slight crest

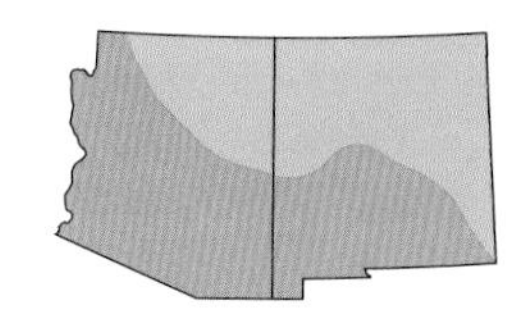

Behavior

With the largest range of the North American tanagers, the Summer Tanager nests in much of Arizona and New Mexico. Snags bees and wasps in midair, sometimes even raiding their hives. After catching one, brings the prey back to its perch, then beats it against a branch and wipes the body along bark to remove the stinger before eating. Also feeds on insects, caterpillars, and fruit it deliberately and methodically picks from leaves. Melodic, warbling song is robin-like. Call is a staccato *ki-ti-tuk*.

Habitat

In Arizona and New Mexico, found in riparian areas with cottonwoods, and in lower canyons.

Local Sites

Very common along Sonoita Creek near Patagonia and along the San Pedro River and the Rio Grande. A somewhat rare migrant across the north.

FIELD NOTES Hepatic Tanager, *Piranga flava* (inset: female left, male right), is a common summer resident in coniferous forests and oak hillsides in southeastern Arizona and southwestern New Mexico. The male has a darker, more brick-red coloration than the Summer Tanager; close up both sexes have gray bills and a dark ear patches.

Breeding | Adult male

WESTERN TANAGER

Piranga ludoviciana L 7¼" (18 cm)

FIELD MARKS

Bright red hood on breeding male

Yellow underparts, nape, and rump; yellow-green face on female

Black wings and tail; upper wing bar yellow; lower wing bar white

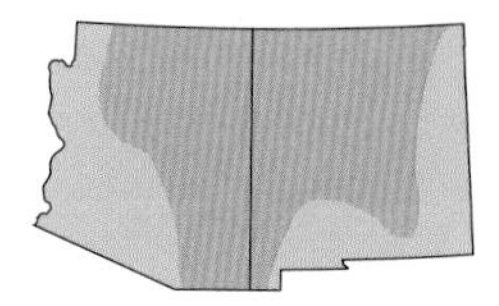

Behavior

Forages mostly in trees for insects, especially wasps and bees, and for fruit. May join mixed-species foraging flocks after breeding when males lose their red hoods for the winter months. Known to visit birdbaths, but rarely feeders. Both sexes are known to sing a hoarse three- to five-phrase series of repeated *chu-wees*. Calls sounding like *pit-ick*, *pit-er-ick*, or *tu-weep* are given year-round.

Habitat

The Western Tanager is found in coniferous and pine-oak forests. Cup-shaped nest is located far out on branches.

Local Sites

Commonly seen high up in both the Santa Catalina and Chiricahua Mountains in Arizona, and in the Sangre de Cristo and Sandia Mountains of New Mexico. Also a common migrant throughout, especially along riparian areas.

FIELD NOTES The Flame-colored Tanager, *Piranga bidentata* (inset: male), is a species that is very rare in the spring and summer in the mountains of southeast Arizona. It has bred in pure pairs and hybridized with Western Tanagers. Check with other birders or telephone hotlines for any current sightings.

Year-round | Adult male

SPOTTED TOWHEE

Pipilo maculatus L 7½" (19 cm)

FIELD MARKS

Black upperparts and hood on male; gray-brown on female

White underparts; white spots on back and scapulars; two white wing bars; rufous sides

Long tail with white corners

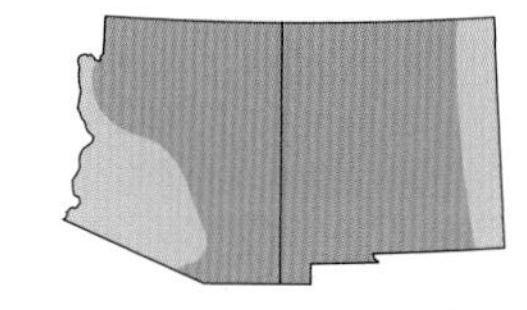

Behavior

This towhee employs its signature double-scratch as it kicks its feet among leaf litter, head held low and tail pointed up, attempting to expose seeds, fruit, and small arthropods, especially beetles, caterpillars, and spiders. Generally seen singly or in a pair, but family groups may stay together for a short time after nesting. Sings a simple trill of variable speed from an exposed perch, though geographical variations occur. Call is an upslurred, inquisitive *queee.*

Habitat

The Spotted Towhee is common in chaparral, brushy thickets, and forest edges. Nests on ground, and occasionally in low trees or shrubs.

Local Sites

Common permanent resident of lower canyon hillsides and chaparral throughout most of the major mountain ranges in Arizona and New Mexico. Generally winters at lower elevations,

---

FIELD NOTES Once classified as the same species as the Eastern Towhee, *Pipilo erythrophthalmus,* the Spotted is now considered a separate species. It is distinguished from the Eastern Towhee by its vocalizations, and by white spotting on its upperparts.

Year-round | Adult

CANYON TOWHEE

Pipilo fuscus L 8" (20 cm)

FIELD MARKS

Grayish overall; reddish crown

Buffy eye ring

Pale throat bordered by streaks

Large whitish belly patch with diffuse dark breast spot

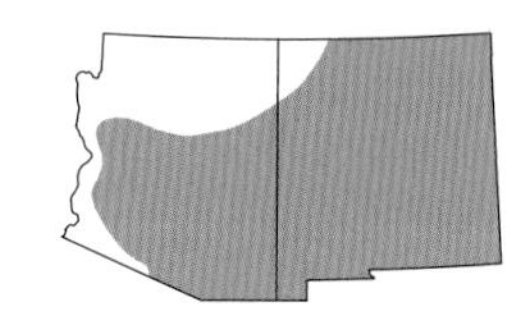

Behavior

Uses the "double scratch" foraging technique: with head low and tail up in the air, the bird rakes both feet over the ground, exposing seeds and beetles. Sometimes flies after insects. Call is a shrill *chee-yep* or *chedup*. Song opens with a call note, which is followed by sweet, slurred notes.

Habitat

Arid, hilly country; desert canyons. Female builds large open-cup nest on the ground or in lower branches of shrubs in dense underbrush.

Local Sites

Common widespread permanent resident in desert areas at low elevations across much of Arizona and New Mexico. Most common in the southern portion of both states.

FIELD NOTES The Green-tailed Towhee, *Pipilo chlorurus* (inset), differs from the Canyon by its more distinct red cap, a white throat bordered by a dark stripe and a white stripe and olive green upperparts and tail.

Year-round | Adult

ABERT'S TOWHEE

Pipilo aberti L 9½" (24 cm)

FIELD MARKS

Uniform chocolate brown above and below

Pale bill framed by black face

Cinnamon undertail coverts

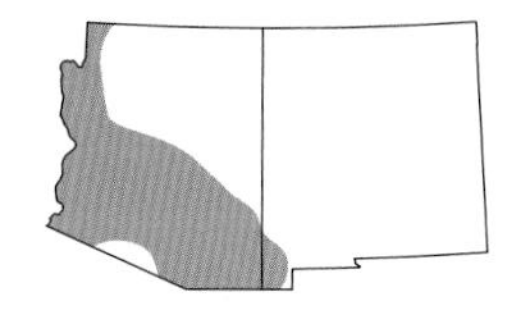

Behavior

The large Abert's Towhee forages by raking the ground with both feet to expose insects and seeds. Its call is a sharp *peek;* song is a series of *peek* notes.

Habitat

Frequents desert woodlands and streamside thickets at lower elevations than the Canyon Towhee. Also found in suburban yards and orchards. Staying in its breeding range year round, it is monogamous and usually mates for life. Female lays three or four pale blue-green eggs in a cup nest close to the ground in a bush or tree.

Local Sites

Common permanent resident in lowland riparian areas and brushy areas along desert washes across much of southern Arizona and extreme southwestern New Mexico. Very common along the Colorado, Salt, Santa Cruz, and San Pedro Rivers in Arizona, and local at San Simone Cienaga in New Mexico.

FIELD NOTES This uniquely patterned towhee, with uniform brown plumage and a black face, is often secretive. Pairs, or small groups, often stay in contact with each other with a distinctive duet call.

Breeding | Adult

CHIPPING SPARROW

Spizella passerina L 5½" (14 cm)

FIELD MARKS

Streaked brown wings and back; unstreaked gray breast and belly; dark line through eye

Breeding adult has chestnut crown; gray cheek and nape

Winter adult has streaked brown crown and a brown face

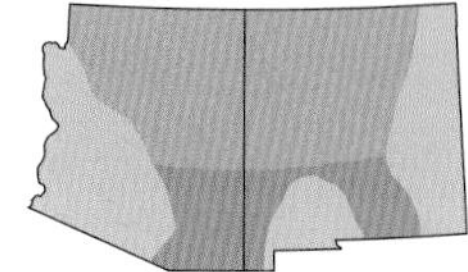

Behavior

Forages on the ground for insects, caterpillars, spiders, and seeds. May be found foraging in small family flocks in fall or in mixed-species groups in winter. Sings from high perch a one-pitched, rapid-fire trill of dry *chip* notes. Call in flight or when foraging is a high, hard *seep* or *tsik.*

Habitat

The Chipping Sparrow can be found in suburban lawns and gardens, woodland edges, and pine and oak forests. Frequents more open areas in winter. Nests close to the ground in branches or vine tangles.

Local Sites

Common summer resident in open coniferous forests and montane foothills throughout major mountains of Arizona and New Mexico. Common winter resident throughout much of southern Arizona and New Mexico.

FIELD NOTES Brewer's Sparrow, *Spizella breweri* (inset), a common breeding species in sagebrush habitat across much of northern Arizona and New Mexico, and common winter resident in brushy edges of agricultural areas in the south, has a brown crown with five streaks and a prominent white eye ring.

Year-round | Adults

LARK SPARROW

Chondestes grammacus L 6½" (17 cm)

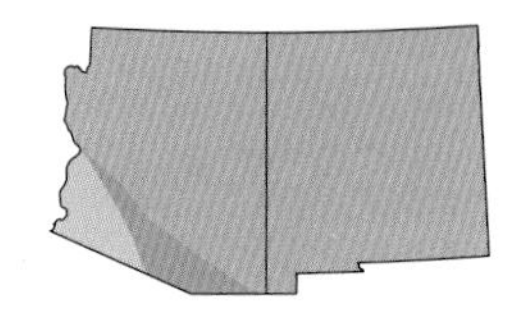

FIELD MARKS

Bold head pattern of chestnut, white, and black

Whitish underparts; dark central breast spot; buffy flanks

Back streaked brown and black

White corners on rounded tail

Behavior

Forages in flocks for seeds, insects, and caterpillars, either on ground or in low branches. During courtship, male swaggers on ground and spreads his tail to show off white feathers. A frequent singer, the Lark Sparrow vocalizes on ground, from perch, while flying, and even at night. Song begins with two loud, clear notes, followed by a series of rich, melodious notes and trills, then unmusical buzzes. Call is a metallic *tsip*, mostly heard in flight.

Habitat

For breeding prefers grasslands, roadsides, farms, grassy woodlands, and orchards. Nests on grass, or low in bush or tree. During winter often assembles in flocks in agricultural areas, semiarid grasslands, and larger lowland parks and fields.

Local Sites

Fairly common summer resident in brushy desert areas with scattered trees; winters commonly in open brushy desert and in agricultural fields.

FIELD NOTES Uniquely patterned face makes this a very distinctive sparrow. Forms single-species flocks that often sit along fence lines. Can easily be coaxed into the open by making pishing sounds.

Year-round | Adult

BLACK-THROATED SPARROW

Amphispiza bilineata L 5½" (14 cm)

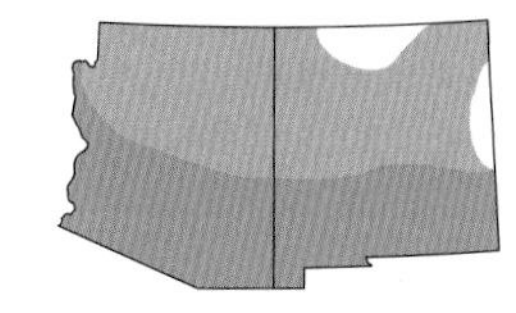

FIELD MARKS

Black lores, chin, throat, and upper breast contrast with white eyebrows and malar stripes

Rounded black tail has white corners that show in flight

Juvenile has white throat; streaked breast

Behavior

Usually paired or in a small group, forages for seeds, spiders, insects, and plant shoots on ground or low in vegetation. Often walks or runs with tail cocked. Nesting times vary from year to year, dependent upon amount of rainfall, hence availability of food. Generally raises two broods a year. Song begins with two clear notes, followed by rapid high-pitched trill, heard often at dawn. Calls are faint, high-pitched, tinkling notes.

Habitat

Inhabits arid and semiarid desert scrublands and rocky slopes. Female builds nest in thorny bush or cactus, using grasses, plant fibers, and hair.

Local Sites

Common permanent resident of desert scrub habitat across the lowlands of southern Arizona and New Mexico. Less common and local in summer across much of the northern portion of both states. Very common in desert around Phoenix and Tucson.

FIELD NOTES The Black-chinned Sparrow, *Spizella atrogularis* (inset: juvenile left, male right), is similar in name but looks very different. The Black-chinned lacks the large black throat patch, has no white on its head, and has rusty scapulars with black streaks on the back.

Winter | Male

LARK BUNTING

Calamospiza melanocorys L 7" (18 cm)

FIELD MARKS

Heavy blue-gray bill

White or buff wing patches

Breeding male has black body; female is brown above, streaked below; winter male like female but blacker face and wings

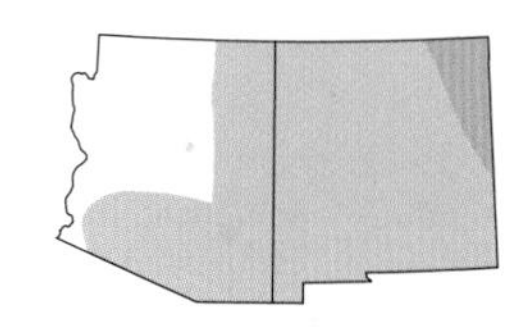

Behavior

In winter, Lark Buntings are often seen in large flocks. They feed on the ground, foraging for seeds and insects. Small numbers nest in northern New Mexico, but the majority breed outside our region, on dry western prairies as far north as Canada. They are early fall migrants, arriving in the southern wintering areas as early as late July and remaining until May. The distinctive call is a soft *hoo-ee,* and the song is a varied series of rich whistles and trills.

Habitat

In winter, favors valley floors, especially those with grasses and weeds, and a variety of other open habitats. Breeding habitat is mostly native shortgrass prairie.

Local Sites

Uncommon and local summer resident in extreme northeastern New Mexico. Common to abundant winter resident, particularly in Sulphur Springs Valley.

FIELD NOTES In breeding plumage, male Lark Buntings (inset) are striking black birds with large white wing patches. But winter males are very different, with mottled and streaked bodies (see photograph). By early May, before leaving the southern parts of our region, the males molt back into their black-and-white plumage.

Year-round | Adult

SAVANNAH SPARROW

Passerculus sandwichensis L 5½" (14 cm)

FIELD MARKS

Yellow or whitish eyebrow

Pale median crown stripe

Strong postocular stripe

Variable streaked upperparts and underparts

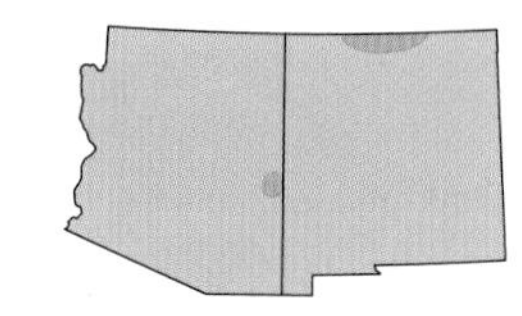

Behavior

Forages on the ground for insects, spiders, and sometimes snails in the summer, seeds and berries in the winter. The Savannah Sparrow is a strong flier and usually not difficult to see—often flying up from the ground and alighting in a small tree or bush. Song begins with two or three *chip* notes, then two buzzy trills. Distinctive flight call is a thin *seep*.

Habitat

Common in a variety of open habitats, marshes, and grasslands. Nests in ground depressions or self-made scrapes sheltered by vines or tall grasses.

Local Sites

Uncommon local summer resident in the White Mountains of eastern Arizona and extreme northern New Mexico. Common, sometimes abundant winter resident in grasslands and agricultural fields across southern Arizona and New Mexico. Very common in the Sulphur Springs Valley.

FIELD NOTES The similar Vesper Sparrow, *Pooecetes gramineus* (inset), nests in sagebrush habitat across northern Arizona and New Mexico, and winters commonly in the same habitat as the Savannah. The Vesper has white outer tail feathers and a distinctive white eye ring.

Year-round | Adult

SONG SPARROW

Melospiza melodia L 5¾" (16 cm)

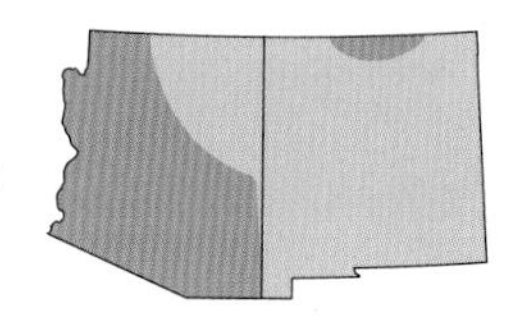

FIELD MARKS

Upperparts streaked

Underparts whitish, with coarse streaking and central breast spot

Long, rounded tail

Broad, grayish eyebrows; broad, dark malar stripes

Behavior

An extremely widespread and variable sparrow, occurring coast to coast and far to the north. Pumps tail up and down in flight. Scratches ground with feet to unearth grain, seeds, berries, and insects. Also forages in trees and bushes and on the ground for larvae, fruits, and berries. Typical song, though variable, is three to four short, clear notes followed by a buzzy *tow-wee* and a trill. Distinctive call is a nasal, hollow *chimp.*

Habitat

Found in the understory along streams and rivers. Winters in a variety of habitats, including brush borders and open fields.

Local Sites

The Song Sparrows numbers are augmented during migration and winter, when it can be seen virtually anywhere in both states. Very common along streams and rivers such as the San Pedro River.

FIELD NOTES Lincoln's Sparrow, *Melospiza lincolnii* (inset), a common winter species across much of southern Arizona and New Mexico, has finer streaks on its buffy breast, a different head pattern, and very different calls.

Year-round | Adult "Gambel's"

WHITE-CROWNED SPARROW

Zonotrichia leucophrys L 7" (18 cm)

FIELD MARKS

Black-and-white striped crown

Underparts mostly gray

Brownish upperparts with blackish brown streaking

Pink, orange, or yellowish bill—depending on subspecies

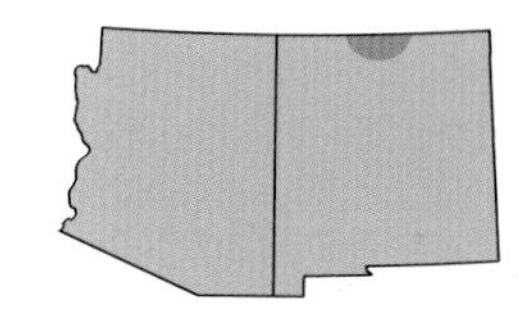

Behavior

Scratches feet along the ground, foraging for insects, caterpillars, and seeds. The operation is audible in areas where wintering flocks congregate. Also gleans food from vegetation. Song variable by region and often heard in winter. Usually one or more thin, whistled notes followed by a twittering trill. Calls include a loud *pink* and a sharp *tseep*.

Habitat

The White-crowned Sparrow occurs in woodlands, grasslands, roadside hedges. Nests on piles of grass or moss, usually in a bush or tree.

Local Sites

Very common to locally abundant migrant and winter resident. Local breeder in the Sangre de Cristo Mountains in northern New Mexico.

FIELD NOTES The common winter subspecies, "Gambel's" White-crowned has whitish supraloral areas and an orange bill. The form that breeds in northern New Mexico and passes through Arizona as a migrant, however, has dark supraloral areas and a dark pink bill.

Year-round | Adult "Gray-headed"

DARK-EYED JUNCO

Junco hyemalis L 6¼" (16 cm)

FIELD MARKS

"Gray-headed" described below

Mostly gray plumage; white outer tail feathers visible in flight; rufous back

Pink bill; black face; dark eyes

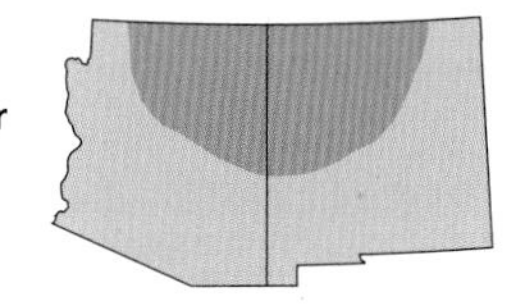

Behavior

Scratches on ground and forages by gleaning seeds, grain, berries, insects, caterpillars, and fruit from plants. Occasionally gives chase to a flying insect. Forms flocks in winter, when males may remain farther north or at greater elevations than juveniles and females. Song is a short, musical trill that varies in pitch and tempo. Calls include a sharp *dit,* and a rapid twittering in flight.

Habitat

Summers in coniferous forests. Winters in a variety of habitats, often in patchy wooded areas.

Local Sites

The "Gray-headed" and "Red-backed" (with dark upper mandible) subspecies are common summer residents in coniferous forests at higher elevations in most of the region's northern mountains. "Gray-headed" winters commonly across southern Arizona and New Mexico. "Red-backed" Juncos breed at the Grand Canyon and across the Mogollon Rim into New Mexico.

FIELD NOTES The Yellow-eyed Junco, *Junco phaeonotus* (inset), commonly breeds in the mountains of southeastern Arizona and extreme southwestern New Mexico. It has a similar red-brown back but can be distinguished from the Dark-eyed Junco by its bright yellow eyes.

Year-round | Adult male

NORTHERN CARDINAL

Cardinalis cardinalis L 8¾" (22 cm)

FIELD MARKS

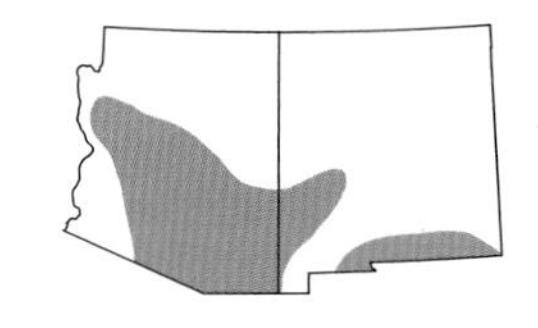

Conspicuous crest

Cone-shaped reddish bill

Male red overall; black face

Female buffy brown tinged with red on wings, crest, and tail

Juvenile browner; dusky bill

Behavior

Forages on the ground or low in shrubs for a wide variety of insects, but mainly feeds on seeds, leaf buds, berries, and fruit. Readily visits backyard feeders and prefers sunflower seeds. Aggressive in defending its territory, the Northern Cardinal will attack not only other birds, but also itself, reflected in windows, rear-view mirrors, chrome surfaces, and hubcaps. Sings a variety of melodious songs year-round, including a *cue-cue-cue*, a *cheer-cheer-cheer*, and a *purty-purty-purty*. Common call is a sharp *chip*.

Habitat

Year-round resident in suburban gardens, woodland edges, streamside thickets, and practically any environment that provides thick, brushy cover for feeding and nesting.

Local Sites

Look for it in New Mexico along the Gila River, and at Rattlesnake Springs. and in Arizona, in riparian areas such as the San Pedro, Salt, and Gila Rivers.

FIELD NOTES Cardinals are monogamous and apparently mate for life—staying together year-round. With the increase of song activity in spring, both birds engage in singing. Males sing loudly and often; the female's song is softer and less frequent.

Year-round | Adult male

PYRRHULOXIA

Cardinalis sinuatus L 8¾" (22 cm)

FIELD MARKS

Stubby yellow bill, with strongly curved upper mandible

Male gray overall; red on face, crest, wings, tail and underparts

Female shows little or no red

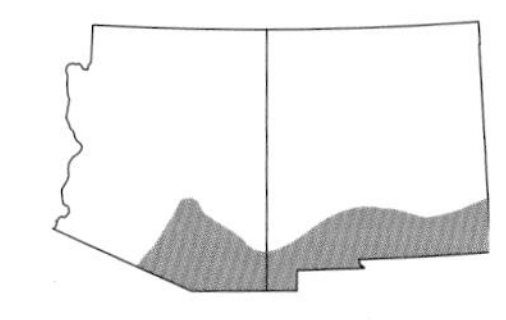

Behavior

Feeds on seeds and insects. Uses strong bill to crush mesquite beans. Song is a liquid whistle, thinner and shorter than the Northern Cardinal's; call is a sharper *chink*. Shy and secretive, the Pyrrhuloxia responds to pishing noises made by a birder. Often found in small flocks after breeding season.

Habitat

Moves in small groups through thorny brush, mesquite thickets, woodland edges, ranchlands. Nest is a loosely built cup of grass, twigs, and bark strips hidden in dense, thorny brush. Female lays three or four speckled white eggs.

Local Sites

Fairly common permanent resident in lush desert washes across much of southeastern Arizona and southern New Mexico. Easily seen around the Arizona Sonoran Desert Museum and Saguaro National Park near Tucson.

FIELD NOTES Thick, strongly curved, pale bill helps distinguish this species from female and juvenile Northern Cardinal. During the nonbreeding season, Pyrrhuloxias tend to wander north and east. Name comes from Latin and Greek words meaning "bullfinch with a crooked bill."

Breeding | Adult male

BLACK-HEADED GROSBEAK

Pheucticus melanocephalus L 8¼" (21 cm)

FIELD MARKS

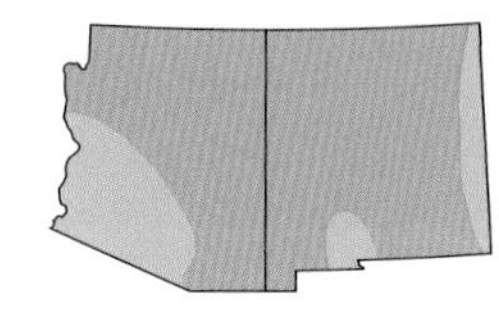

Very large, dark, triangular bill

Yellow wing linings show in flight

Male has cinnamon underparts; black head and upperparts

Female buffy above and below, with little streaking on underparts

Behavior

Forages for seeds, insects, caterpillars, berries, and fruit on the ground and in trees and bushes. In flight, beats wings rapidly in between brief periods spent gliding with wings at sides. Young brooded by both male and female. Rich back-and-forth whistled song reminiscent of American Robin. Call is a high sharp *eek.*

Habitat

Lives in open and streamside woodlands, especially pine and oak, and along forest edges. Nests moderately high up in dense vegetation, usually near water.

Local Sites

Common summer resident of lower montane canyons and coniferous forests. Quite numerous in riparian areas of canyons in southeastern Arizona, such as Madera, Garden, and Cave Creek. A common migrant in riparian habitats throughout both states.

FIELD NOTES Black-headed and Rose-breasted Grosbeaks, *Pheucticus ludovicianus,* are very closely related. The two species occasionally interbreed where their ranges overlap in the Great Plains. The resulting hybrids are very tricky to identify, especially females and immatures.

Year-round | Adult male

BLUE GROSBEAK

Passerina caerulea L 6¾" (17 cm)

FIELD MARKS

Very large bill; chestnut wing bars; chunky body

Male mostly blue; black face

Female warm brown; paler below

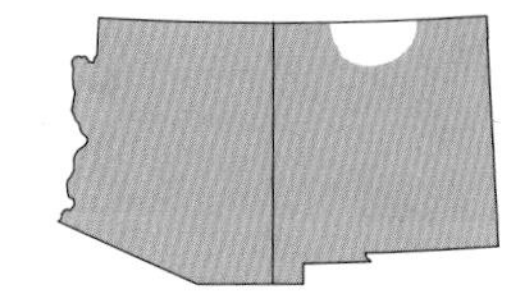

Behavior

Often seen in groups of two to five during migration. Forages for seeds and insects low in weeds. Often twitches and spreads its tail. Call is a loud, explosive *chink.* Song is a series of rich, rising and falling warbles.

Habitat

The Blue Grosbeak prefers low, overgrown fields, streamsides, woodland edges, brushy roadsides. Males sing from high perches to attract females and proclaim territory. Females build small bowl-shaped nest, usually low in small trees, shrubs, or tangles.

Local Sites

Common summer resident of lowland riparian areas throughout the region. Commonly seen along small streams such as Sonoita Creek near Patagonia, and along rivers such as the San Pedro, Rio Grande, and Pecos. Arrives on breeding grounds in May.

FIELD NOTES Sometimes confused with the Indigo Bunting, *Passerina cyanea* (inset; male), an uncommon migrant and local summer resident in both Arizona and New Mexico. The Indigo lacks the black face and chestnut wing bars of the Blue Grosbeak.

Year-round | Adult male

LAZULI BUNTING

Passerina amoena L 5½" (14 cm)

FIELD MARKS

Adult male is bright turquoise above and on throat; cinnamon across breast; thick white wing bars

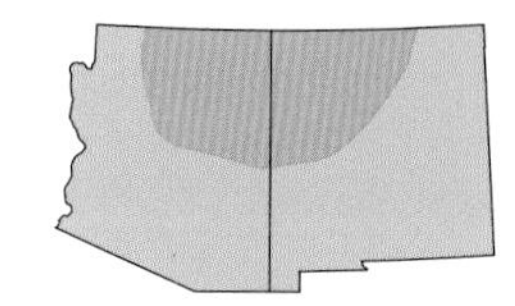

Female is grayish brown above; grayish blue rump and buffy underparts

Behavior

Eats insects and seeds. Male's courtship display involves fluttering on the ground with extended trembling wings. Song is a series of varied phrases, sometimes paired. It is faster and less strident than the Indigo Bunting's song. Call is a dry, sharp *pit.*

Habitat

The Lazuli Bunting favors open deciduous or mixed woodlands, or chaparral, especially in brushy areas near water. Builds nest of coarsely woven dried grass in thickets or shrubby growth, usually near water.

Local Sites

Fairly common local summer resident in riparian areas across the Mogollon Rim country of Arizona into central New Mexico. A common migrant virtually anywhere in both states. Winters in small numbers in the Nogales-Patagonia area of southern Arizona.

FIELD NOTES The Varied Bunting, *Passerina versicolor* (inset: male), is a closely related species found in more arid habitats—desert washes and canyon hillsides. In poor light the males appear uniformly dark; in good light they are stunning. Females are very plain, uniform gray-brown.

Year-round | Adult male

RED-WINGED BLACKBIRD

Agelaius phoeniceus L 8¾" (22 cm)

FIELD MARKS

Male is glossy black; bright red shoulder patches broadly edged in buffy yellow

Female densely streaked overall

Pointed black bill

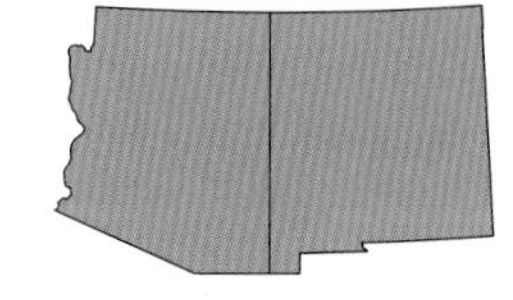

Behavior

Runs and hops while foraging for insects, seeds, and grains in pastures and open fields. The male's bright red shoulder patches are usually visible when it sings from a perch, often atop a cattail or tall weed stalk, defending its territory. At other times only the yellow border may be visible. Territorially aggressive, a male's social status is dependent on the amount of red he displays on his shoulders. Song is a liquid, gurgling *konk-la-reee,* ending in a trill. Call is a low *chack* note.

Habitat

Breeds in colonies, mainly in freshwater marshes and wet fields with thick vegetation.

Local Sites

Marshes and wet fields around lakes and ponds throughout both Arizona and New Mexico. Numbers augmented during migration and winter, when large single-species flocks can be found in agricultural areas.

FIELD NOTES Brewer's Blackbird, *Euphagus cyanocephalus* (inset: female left, male right), breeds across northern Arizona and New Mexico and winters commonly in agricultural areas. The male lacks the Red-winged's red shoulder patches; female is uniformly gray-brown, without streaks.

Breeding | Adult

EASTERN MEADOWLARK

Sturnella magna L 9½" (24 cm)

FIELD MARKS

Yellow below, with black V-shaped breast band, paler in winter

Black-and-white-striped crown; yellow supraloral area

Brown above, streaked with black

White outer tail feathers in flight

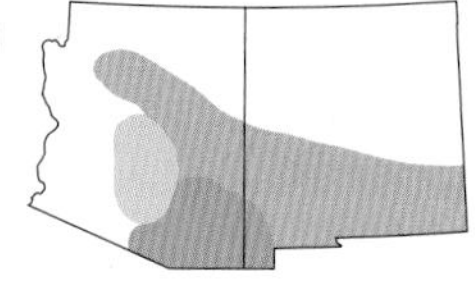

Behavior

Flicks its tail open and shut while foraging on the ground, feeding mainly on insects during spring and summer, seeds and agricultural grain in late fall and winter. Generally solitary in summer, the meadowlark forms small flocks in fall and winter. Often perches on fence posts or telephone poles to sing a clear, whistled *see-you see-yeeer.* Flight call is a buzzy *drzzt.*

Habitat

Prefers open space offered by grasslands, pastures, meadows, and farm fields. Female constructs a domed nest on ground, woven into the surrounding grasses.

Local Sites

The "Lilian's" form of Eastern Meadowlark is the common breeding meadowlark across much of southeastern Arizona and southern New Mexico. Very common in the grasslands.

FIELD NOTES The Western Meadowlark, *Sturnella neglecta* (inset), which winters in southeastern Arizona, looks much like the Eastern. The Western has a different song, more extensively yellow throat, and less white in its tail than the "Lilian's" form of the Eastern Meadowlark.

Year-round | Adult male

YELLOW-HEADED BLACKBIRD

Xanthocephalus xanthocephalus L 9½" (24 cm)

FIELD MARKS

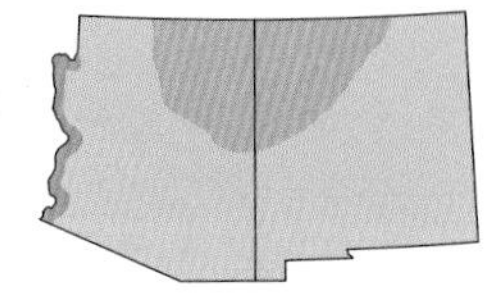

Male has prominent yellow hood and breast; black body and lores; large white wing patch

Female is washed with yellow on her face and breast; dark brown body and crown

Stubby, black, triangular bill

Behavior

Highly gregarious this bird breeds in colonies and forms large flocks outside of the breeding season, sometimes numbering in the thousands. Forages communally on ground, for insects, larvae, snails, grain, and seeds. Highly territorial, the Yellow-headed will attack other birds and even humans who intrude on its territory. Unmusical, raspy song begins with a few loud, harsh notes and ends in a long, descending buzz resembling the sound of a chainsaw. Call note is a hoarse croak.

Habitat

Found in summer in cattails around lakes and ponds. Winters in agricultural fields and roosts in marshy borders of ponds and feedlots.

Local Sites

Common summer resident around marshes and edges of lakes across northern Arizona and New Mexico. Common to abundant winter resident near ponds and in agricultural areas across the southern part of both states.

FIELD NOTES Huge numbers of single-species flocks of Yellow-headeds can be found roosting in marshes, or at feedlots, in southern Arizona. Most winter south into northern Mexico, and form single-sex flocks, sometimes numbering in the millions.

Year-round | Adult male

GREAT-TAILED GRACKLE

Quiscalus mexicanus Male L 18" (46 cm) Female L 15" (38 cm)

FIELD MARKS

Large with very long tail; yellow eyes, duller on female

Male iridescent black overall; purple sheen on head and back

Female dark brown overall, paler on throat and breast

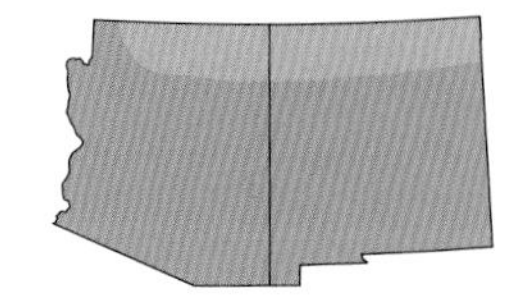

Behavior

In winter, may join large groups numbering in the thousands. Not a territorial bird, flocks forage together, locating rich food sources and sharing their finds. Individuals spend less time on alert for predators, since all glance around for danger. Male's courtship pose consists of wings drooped, tail spread, and bill pointed skyward. Dominant males gather harems of females, with whom they mate and defend against advances by other males. Voices are harsh, with varied calls, including clear whistles and loud *clack* notes. Courtship wing shaking produces strange rattling.

Habitat

Common and increasing in open flatlands with scattered groves of trees, and in marshes and wetlands.

Local Sites

Common to abundant permanent resident of towns, cities, and agricultural areas throughout Arizona and New Mexico. Very common in suburban Phoenix and Tucson.

FIELD NOTES The breeding range of Great-tailed Grackles has expanded north and west during the past 20 years. It now nests across northern Arizona and New Mexico—a relatively recent expansion. It is abundant around towns all through Mexico and Central America.

Year-round | Adult male

BROWN-HEADED COWBIRD

Molothrus ater L 7½" (19 cm)

FIELD MARKS

Pointed bill

In male, brown head and metallic black body with green gloss

Female gray-brown above, paler below

Strong, direct flight

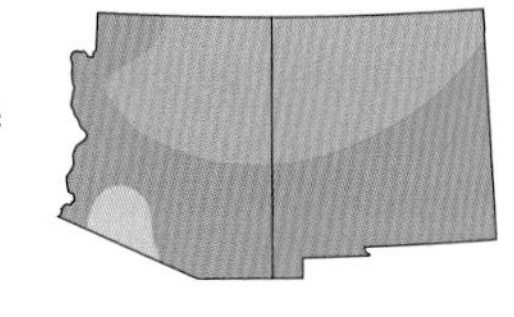

Behavior

Often forages on the ground among herds of cattle, feeding on insects flushed by grazing. Also feeds heavily on grass seeds and agricultural grain, and is sometimes viewed as a threat to crops. Generally cocks its tail up while feeding. Song is a squeaky, gurgling call that includes a squeaky whistle. Cowbirds are brood parasites, laying their eggs in the nests of other species, leaving the responsibilities of feeding and fledging of their young to the host birds.

Habitat

A cowbird prefers the open habitat provided by farmlands, pastures, prairies, and edgelands bordering woods and forests. Also frequents suburbs.

Local Sites

Common summer resident across virtually all of New Mexico and Arizona. Northern birds migrate south for winter. Abundant at cattle feedlots in southwest Phoenix and near Red Rock.

FIELD NOTES The male Bronzed Cowbird, *Molothrus aeneus* (inset), has a larger bill, red eyes and a ruffed neck, and lacks the metallic green sheen of the Brown-headed. It is not nearly as abundant. Both species are parasitic and lay eggs in other species' nests.

Breeding | Adult male

HOODED ORIOLE

Icterus cucullatus L 8" (20 cm)

FIELD MARKS

Breeding male orange-yellow with black patch on throat; mostly black wings with two white bars; long black tail

Female has greenish gray upperparts; yellowish under-parts; two wing bars

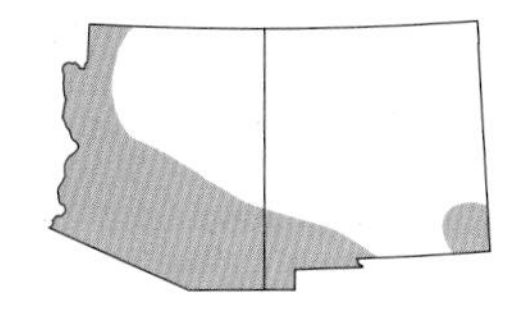

Behavior

The Hooded Oriole forages in trees for a variety of insects and uses pointed bill to draw nectar from flowers. Also eats fruit, visits bird feeders, and drinks from hummingbird feeders. Male may initially chase after female, and she may respond by drooping a wing. Eventually, male perches near female and bows repeatedly, showing off plumage. Calls include a whistled, rising *wheet.* Song is a series of whistles, trills, and rattles.

Habitat

Frequents riparian woodland, Sonoran Desert, mesquite, and scrub. Nest is a pouch intricately woven of grasses hung from branches too weak to accommodate predators.

Local Sites

Common summer resident of lowland desert hillsides, particularly in the Sonoran Desert and in riparian areas in southern New Mexico such as Rattlesnake Springs.

FIELD NOTES Bullock's Oriole, *Icterus bullockii* (inset: female left, male right), breeds commonly in riparian areas across the region. Males are deep orange with black cap and eye stripe, white wing patch; females are grayish with yellow head and breast.

Year-round | Adult male

HOUSE FINCH

Carpodacus mexicanus L 6" (15 cm)

FIELD MARKS

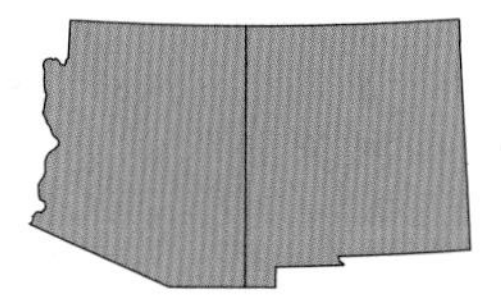

Male's forehead, bib, and rump typically red, but can be orange or yellow

Brown streaked back and flanks

Female streaked dusky brown on entire body

Behavior

A seed eater, the House Finch forages on the ground, in fields and in suburban yards. Often visits backyard feeders. Seen in large flocks during winter. Flies in undulating pattern, during which squared-off tail is evident. Male sings a conspicuously lively, high-pitched song consisting of varied three-note phrases, usually ending in a nasal *wheer.* Calls include a whistled *wheat.*

Habitat

Adaptable to varied habitats, this abundant bird prefers open areas, including suburban parks and areas where it can build its cup-like nest on buildings. Also nests in shrubs, trees, cactuses, or on the ground.

Local Sites

Common permanent resident in a variety of habitats across virtually all of Arizona and New Mexico. Less common in the far north during winter. Very common around towns and cities. Often visits feeders.

FIELD NOTES The Pine Siskin, *Carduelis pinus* (inset), a common summer resident in pine forests across many of the mountain ranges and a common winter species, locally, in lowland areas of the south, looks similar but has a more pointed bill and a yellow flash in its primaries.

Year-round | Adult male "Green-backed"

LESSER GOLDFINCH

Carduelis psaltria L 4½" (11 cm)

FIELD MARKS

Male black on hood, back, and tail; bright yellow below

Female greenish olive above, dull yellow below

White wing bars and edges to tertial and primary flight feathers

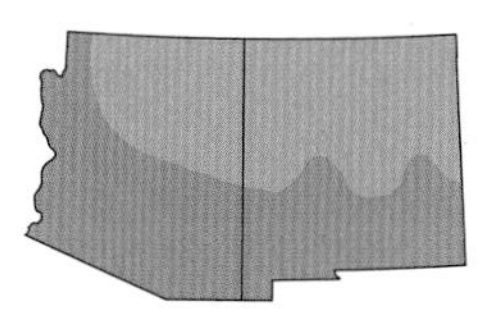

Behavior

Pairs or small flocks forage for seeds and insects in bushes, shrubs, and weedy fields. Commonly visits birdbaths and outdoor faucets in the semiarid southwest, as diet primarily of seeds does not provide a lot of moisture. Male feeds female while she is brooding, and pairs may stay together for life. Song, given by male in flight, is a lively series of warbles and *swee* notes. Call, given often by small flocks, is a *tee-yee tee-yer.*

Habitat

Found in dry brushlands and open woodlands with scattered trees. Also tends to areas of human habitation to take advantage of artificial sources of water. Female builds nest in bushes or trees, sometimes in tall weeds.

Local Sites

Common summer resident in riparian areas. Common winter resident across southern Arizona and New Mexico. Easily seen during winter in riparian areas such as Sonoita Creek.

FIELD NOTES The American Goldfinch, *Carduelis tristis* (inset: winter male), a less common wintering species across much of southern Arizona and New Mexico, looks similar but the winter male and female American is best distinguished from its smaller cousin by its white (not yellow) undertail coverts.

Summer | Adult male

HOUSE SPARROW

Passer domesticus L 6¼" (16 cm)

FIELD MARKS

Summer male has black bill, bib, and lores; chestnut eye stripes and nape

Winter male has chestnut and black areas veiled with gray

Female brown with streaked back; buffy eye stripe

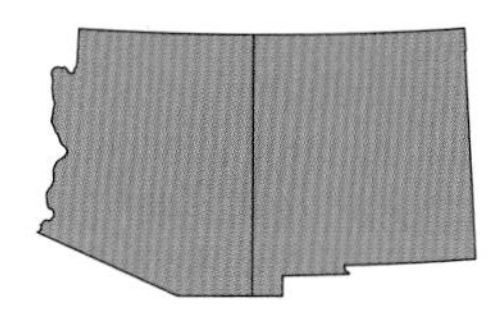

Behavior

The House Sparrow is abundant and gregarious year-round. Hops around, feeding on grain, seeds, and shoots, or seeks out bird feeders for sunflower seeds and millet. In urban areas, begs for food from humans and will clean up any crumbs left behind. In spring and summer, multiple suitors will chase a possible mate. Females choose mate mostly according to song display. Singing males give persistent *cheep*.

Habitat

Found in close proximity to humans. Can be observed in urban and suburban areas and in rural landscapes inhabited by humans and livestock. Nests in any sheltered cavity, often usurping it from another species.

Local Sites

Common to abundant permanent resident in towns and cities across Arizona and New Mexico.

FIELD NOTES These Old World sparrows were originally introduced in eastern North America and have expanded to virtually the entire country. The female (inset) is much plainer then the male; juveniles look similar to the female.

Mostly Black

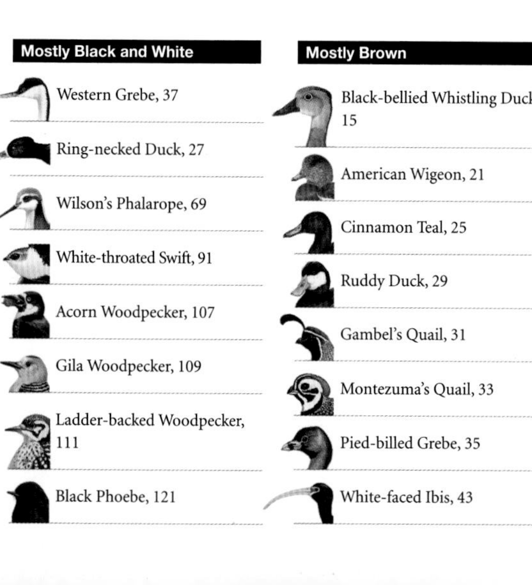

Mostly Black and White

Mostly Brown

Mostly Brown and White

Mostly Gray

Mostly Blue

Mostly Greenish

Prominent Orange Head

Mostly White

Mostly Yellow

Yellow and Black

Prominent Green Head

Prominent Red Head

The main entry page number for each species is listed in **boldface** type and refers to the text page opposite the illustration.

A check-off box is provided next to each common-name entry so that you can use this index as a checklist of the species you have identified.

ACKNOWLEDGMENTS

The Book Division would like to thank the following people for their guidance and contribution in creating the *National Geographic Field Guide to Birds: Arizona & New Mexico*

Tom Vezo:
Tom Vezo is an award-winning wildlife photographer who is widely published throughout the U.S. and Europe. Located out of Green Valley, Arizona, he specializes in bird photography but photographs other wildlife and nature subjects as well. He is also a contributor to the *National Geographic Reference Atlas to the Birds of North America.* For a look at more of his images, find his gallery at tomvezo.com.

Brian E. Small:
Brian E. Small has been a full-time professional wildlife photographer specializing in birds for more than 15 years. In addition, he has been a regular columnist and Advisory Board member for *WildBird* magazine for the past 10 years. An avid naturalist and enthusiastic birder, Brian is currently the Photo Editor for the American Birding Association's *Birding* magazine. You can find more of his images at www.briansmallphoto.com.

Cortez C. Austin, Jr.:
Cortez Austin is a wildlife photographer who specializes in North American and tropical birds. He has a degree in zoology and has done graduate work in conservation, ecology, and microbiology. An ardent conservationist, he has donated images, given lectures, and written book reviews for conservation organizations. In addition he has published numerous articles and photographs in birding magazines in the United States. His photographs have also appeared in field guides, books, and brochures on wildlife.

Bates Littlehales:
National Geographic photographer for more than 30 years covering myriad subjects around the globe, Bates Littlehales continues to specialize in photographing birds and is an expert in capturing their beauty and ephemeral nature. Bates is co-author of the *National Geographic Photographic Field Guide: Birds,* and a contributor to the *National Geographic Reference Atlas to the Birds of North America.*

Larry Sansone:
An active birder since 1960, Larry Sansone began photographing wildlife in the early 1970s. His pictures are published in field guides and magazines in the U.S. and Europe. He was a technical advisor to the first edition of the *National Geographic Field Guide to the Birds of North America,* and he is photo editor of *Rare Birds of California* by the California Bird Records Committee.

ILLUSTRATIONS CREDITS

Photographs

Cortez C. Austin, Jr.: pp. 66, 150, 188; **Richard Crossley:** p. 174; **G.C. Kelley:** cover, pp. 32, 74, 92, 96, 98, 100, 106, 236; **Peter LaTourette:** p. 158; **Bates Littlehales:** pp. 180, 220; **Gary Rosenberg:** pp. 102, 122; **Larry Sansone:** pp. 16, 72, 118, 120, 142, 144, 148, 182, 190, 194, 202, 242, 246; **Brian E. Small:** pp. 26, 42, 44, 50, 60, 76, 78, 80, 86, 90, 104, 110, 114, 116, 124, 126, 128, 130, 132, 134, 136, 152, 156, 162, 166, 168, 170, 172, 178, 186, 192, 196, 200, 204, 208, 210, 218, 224, 226, 228, 234, 248, 254, 258; **Tom Vezo:** pp. 2, 14, 18, 20, 22, 24, 28, 30, 34, 36, 38, 40, 46, 48, 52, 54, 56, 58, 62, 64, 68, 70, 82, 84, 88, 94, 108, 112, 138, 140, 146, 154, 160, 164, 176, 184, 198, 206, 212, 214, 216, 222, 230, 232, 238, 240, 244, 250, 252, 256, 260

NATIONAL GEOGRAPHIC
FIELD GUIDE TO BIRDS:
ARIZONA & NEW MEXICO

Edited by Jonathan Alderfer

Published by the National Geographic Society

John M. Fahey, Jr., *President and Chief Executive Officer*

Gilbert M. Grosvenor, *Chairman of the Board*

Nina D. Hoffman, *Executive Vice President; President, Books and Education Publishing*

Prepared by the Book Division

Kevin Mulroy, *Senior Vice President and Publisher*

Kristin Hanneman, *Illustrations Director*

Marianne R. Koszorus, *Design Director*

Rebecca Hinds, *Managing Editor*

Carl Mehler, *Director of Maps*

Barbara Brownell Grogan, *Executive Editor*

Staff for this Book

Kate Griffin, *Project Manager*

Dan O'Toole, *Illustrations Editor*

Carol Norton, *Art Director, Bird Program*

Alexandra Littlehales, *Designer*

Ann Oman, *Writer*

Suzanne Poole, *Text Editor*

Teresa Tate, *Illustrations Specialist*

Matt Chwastyk, Sven M. Dolling, *Map Production*

Gary Rosenberg, *Map Research*

Lauren Pruneski, *Editorial Assistant*

Rick Wain, *Production Project Manager*

Manufacturing and Quality Control

Christopher A. Liedel, *Chief Financial Officer*

Phillip L. Schlosser, *Managing Director*

John T. Dunn, *Technical Director*

National Geographic Partners
1145 17th Street NW
Washington, DC 20036-4688 USA

ISBN: 978-0-7922-5312-9

Printed in China

16/PPS/3